Homage to a Pied Puzzler

Homage to a Pied Puzzler

Edited by
Ed Pegg Jr.
Alan H. Schoen
Tom Rodgers

CRC Press
Taylor & Francis Group
Boca Raton London New York

CRC Press is an imprint of the
Taylor & Francis Group, an **informa** business

AN A K PETERS BOOK

CRC Press
Taylor & Francis Group
6000 Broken Sound Parkway NW, Suite 300
Boca Raton, FL 33487-2742

First issued in paperback 2017

ISBN-13: 978-1-56881-315-8 (hbk)
ISBN-13: 978-1-138-11501-9 (pbk)

Library of Congress Cataloging-in-Publication Data

Homage to a pied puzzler / edited by Ed Pegg Jr., Alan H. Schoen, Tom
Rodgers.
   p. cm.
  Includes bibliographical references.
  ISBN 978-1-56881-315-8 (alk. paper)
 1. Mathematical recreations. I. Pegg, Ed, 1963- II. Schoen, Alan H. (Alan
Hugh), 1924- III. Rodgers, Tom, 1943-
QA95.H656 2008
793.74–dc22

                2008013455

Visit the Taylor & Francis Web site at
http://www.taylorandfrancis.com

and the CRC Press Web site at
http://www.crcpress.com

Contents

Contents

Preface

Every two years or so, mathematicians, puzzlers, and magicians assemble in Atlanta to pay tribute to Martin Gardner, whose life-long devotion to recreational mathematics and to magic is honored throughout the world. The seventh of these "Gatherings for Gard-ner," G4G7, was held March 16-19, 2006. Most of the articles in this volume are drawn from oral presentations delivered at G4G7. Overflow articles will appear in a companion volume, Mathematical Wizardry for a Gardner.

Mathematicians everywhere recognize that even though Martin Gardner is not himself a professional mathematician, he has been astoundingly effective in popularizing not just recreational mathematics, magic, and puzzles, but also so-called serious mathematics. A number of professional mathematicians have acknowledged that their passion for their subject was first kindled by one of Martin's columns or essays concerning either a hoary old chestnut or a brand-new problem. His columns in Scientific American, which first appeared in 1956, earned him readers worldwide, and he continues writing to this day. His published writings are available in both book form and on CD.

Martin's unique influence has been the result of a refined taste in subject matter combined with a famously clear and witty writing style. His well-known addictions to magic and to word puzzles have inspired countless others to attempt to emulate him in these arts, but he is still the undisputed master. Perhaps his most brilliant invention is Dr. Irving Joshua Matrix, a proxy he used to spoof virtually every known variety of charlatan, including-but not limited to-Biblical cryptographers and numerologists.

Those among us who were fortunate to attend G4G6 will never forget that Jay Marshall, the legendary iconoclast who was loved and admired by all who revere magic and magicians, graced us with his presence. Even though Jay was not at all well, he cheerfully performed one of his classic routines. We were deeply moved. Not many weeks later, Jay shuffled off this mortal coil. The first two articles in this book are dedicated to his memory.

It was Tom Rodgers who in 1993 first thought of assembling admirers of Martin Gardner for a few days of celebration. He has hosted all of these gatherings since then, thanks to the fine work by the co-organizers Elwyn Berlekamp in the math community and Mark Setteducati in the magic world, as well as the generous help of Karen Farrell, Jeremiah Farrell, Scott Hudson, Thane Plambeck, Emily DeWitt Rodgers, Stephen Turner, and the many G4G participants who volunteered at moments of need.

The editors are especially indebted to Charlotte Henderson, associate editor for our publisher, A K Peters, Ltd., for her patient ministrations in the task of assembling and editing authors' contributions.

We hope that readers derive as much pleasure from this volume as we have had in assembling it.

Ed Pegg Jr.
Champaign-Urbana, Illinois

Alan H. Schoen
Carbondale, Illinois

Tom Rodgers
Atlanta, Georgia

In Memoriam

Jay Marshall: A Legacy of Magic and Laughter

Robert Cotner

"Lefty" is silent—and he'll never talk again. You see, his progenitor and master has died.

Born from the creative spirit of Jay Marshall in the South Pacific during World War II—out of a khaki Army glove—Jay and Lefty have traveled the world, entertaining people in London, New York, Las Vegas, Chicago, and all points between. All that ended on May 10, 2005, when Jay Marshall died at Swedish Covenant Hospital in Chicago.

The *Chicago Tribune* (May 13, 2005, Sec. 2, p. 13), in an illustrated obituary, remembered Jay as "Ventriloquist, magic-shop owner" and celebrated his professional career on stage and on television, particularly on the "Ed Sullivan" show, where he and Lefty "gained fame as the opening act for performers ranging from Milton Berle to Liberace." Jay appeared on the Sullivan show 14 times and opened, as well, for Frank Sinatra at the Desert Inn in Las Vegas.

The standard routine between Jay and Lefty included Jay's question, "Shall we sing?"

Figure 1. Jay in 1980 at the Hild branch of the Chicago Public Library, after a Punch and Judy show. Caxtonian Peggy Sullivan discovered this photo in the Special Collections and Preservation division of the library.

Lefty would respond, "What do you want to sing?"

Jay would reply, "If I Had My Way."

And Lefty would counter, "If I had *my* way, I wouldn't sing!" Lefty was, indeed, as the *Tribune* reported, a "wiseacre rabbit."

The article quotes Marc DeSouza, a member of the Society of American Magicians, "Jay is one of the most influential magicians of our age."

Jay came to Chicago in the mid-1950s to marry Frances Ireland, who was also a magician. Frances was the owner of a magic shop that had been in business since the 1930s. She had married the founder in 1931. The founder died in the early 1950s, and sometime after his death, Frances and Jay were married. In 1962, the shop moved to North Lincoln Avenue, where Jay's high-profile role in the circle of American magicians made his shop an institution unrivaled anywhere.

The *Tribune* commented, the "shop in Lincoln Square was a magnet for kids.Once in a while, a boy would declare that when he grew up, he would be a professional magician."

"Son," Jay would tell him, "you can't be both."

Magic Inc. is now run by Jay's grandson, Chris Marshall, until another professional magician can be located to operate it. The *Tribune* reported that there is a wood replica of "Lefty watching over the patrons. There's a fresh tear pasted under his eye."

The *New York Times*, in an illustrated obituary (May 13, p. C13), announced, "Jay Marshall, 85, the Dean of Magic Is Dead. "The article recalled Jay's playing the "Ed Sullivan" show, the New York Palace, and London's Palladium. It reminded us that, since 1992, Jay was the dean of the "Society of American Magicians." He was, the *Times* reported, "A writer, editor, and collector of all things magic, as well as owner of one of the nation's leading stores in the business, Magic Inc."

Calling Jay an "indefatigable performer," the *Times* cited his important role in the closing days of vaudeville, his roles with Paul Robeson, Sid Caesar, and Walter Cronkite, and his origination of a magic trick known as the "Jaspernese Thumb Tie," which "is still the staple of prestidigitators."

The article quotes Siegfried Fischbacher of Siegfried & Roy: "Jay Marshall was a name synonymous with magic. He was one of magic's most beloved figures."

An interviewer from *Genii* reported that Jay, when being interviewed for the deanship of the Society of American Magicians, asked, "What do the dean do?"

The answer given him was, "As far as I know the dean don't do nothin'."

"That's the job for me!" Jay replied.

Jay's Caxton friends remember him differently. David Meyer, perhaps his most intimate friend, has spoken by phone weekly with Jay for the past 25 years and daily for the past three. "He had a great sense of humor," David recalls. One of his non-verbal jokes, David recalled, was to perform with a rubber dove perched on his shoulder. When he turned to walk away from the audience, they saw a ribbon of white bird droppings down the back of his tuxedo jacket. "It broke the audience up!"

Once at dinner, David remembers, someone asked Jay what he thought of English scholar Trevor Hall, who wrote scholarly books on magic, Sherlock Holmes, and Dorothy Sayers, with a notorious stuffiness. Jay, who loved the role of the quick-change artist, leaped to his feet, rushed out of the room, and came back shortly wearing a black academic gown. He took his seat at the table with a dour mien of stiff, haughty formality. "This is Trevor Hall!" Jay announced solemnly to the delighted guests.

The *Times* reported that Jay "could move from mismatched plaids to well-cut evening clothes." David remembers Jay as being up-to-date in dress on stage. "When blue jeans came into fashion," David recalled, "Jay had a tuxedo made of denim."

John McKinvin, now living is Ohio, is a longtime personal friend of Jay and a fellow magician. "Jay was addicted to books," John said. "He would attend every book sale he could, and I used to go with him regularly to these sales." John recalled that Jay was indiscriminate in his buying—and he had so many, many interests, from magic to show business. "He never left a book sale—he especially liked the Lake Forest sales—without at least two shopping bags full of books. I would help him carry the books to his car, and, when he got home, Frances would have a fit!" John remembered. The books are still there, John said. "Jay had to buy some extra buildings just to keep them."

Grandson Chris confirms this: He reported, "There are two two-story buildings full of books and magazines, plus a small house that is partly full of books on the North Lincoln Street property." The family expects that it will take at least six months to sort the collection, and they will determine what to do with it.

John Railing, long-time friend of Jay's and another fellow magician said,"I personally believe that his greatest happiness was from acquiring *books*—not from magic. He was a bibliomaniac in the finest tradition."

John loves telling this "book anecdote" about Jay, related to him by one of his mentors in magic, Harry Riser. John tells the story this way:

I confirmed the details of this story with Jay only last year: When he performed on Broadway (probably as the bagpipe-playing ghost of a butler in the 1951 play, "Great to Be Alive"), Jay was assigned a dressing room in which the wallpaper was of books on a wall-to-wall bookcase. But the spines of the books were blank. So Jay, with his unique, playful mind, occupied his time by filling in the titles and authors of the books on the wallpaper. A very thin book, for example, would be titled, "History of the World," while the thickest spine might be "Origin of the Safety Pin." Once he vacated the dressing room, subsequent actors and actresses followed his lead and continued filling in more of the titles. Unfortunately, sometime later, a "diva" was appalled by the decor of this dressing room and asked for new wallpaper, and we lost what was undoubtedly a very interesting "library."

My own recollection of Jay is probably unique; I remember Jay as the philanthropist. When I was an executive with the Salvation Army in Chicago, he would stop by my office on North Pulaski Road each year and bring a check up to me. If I wasn't in my office, he would leave it at the front desk in an envelope addressed to me.

He was what we called a "major donor," and he loved the work that the Army was doing for less fortunate people. In 1998, I think it was, the Women's Auxiliary planned a charity auction at a club downtown, and they scheduled Ted Amberg, a young magician and the son of Tom Amberg, the Chair of our Board, as entertainment. I called Ted's father and asked if it would make Ted nervous if I brought Jay Marshall with me to meet Ted and watch him perform. "Not at all!" Tom said.

So Jay was my guest for the evening. I picked him up at the shop, and drove to the club for a delightful evening. Jay loved associating with Salvation Army people, both officers and volunteers. Tom recalled the evening recently. "Ted was delighted to meet Jay Marshall. He was an icon in the world of magic, and Ted considered it a high honor to meet him." Ted runs a small entertainment business himself these days, in Springfield, Missouri, where his company was named one of the top five small businesses in Springfield.

John McKinven said that Jay rarely went to men's clothing stores as he got older. "He bought his clothing—at least in his later years—at Salvation Army Thrift Shops. Maybe he didn't want to spend much money—I just don't know," John mused. The answer, I think, is that his buying at Thrift Shops was an extension of his philanthropic spirit: he knew, in buying at Salvation Army shops, he was helping people he cared a great deal about. This was a sure, simple way of making a difference in their lives.

One of the final meetings I had with Jay occurred during the cocktail hour of a Caxton dinner meeting. Jay came up to me and said, "Well, I just came from the Army headquarters." He referred to the Divisional office on Pulaski, where my office used to be. "I asked for the Big Man," he continued. He referred to Lt. Colonel David Grindle, Divisional Commander. "I went into his office, handed him my check, and told him the gift was in honor of Bob Cotner," he concluded.

Figure 2. Publicity photo showing Jay performing handshows, probably taken in the late 1940s. Photograph provided by the Marshall family.

What an honor to be so honored by Jay Marshall! Norma and I sat with Jay at dinner that evening, and we laughed our way through the evening, as was Jay's custom. You always had fun when Jay Marshall was around. That may be his greatest legacy to us, his friends, who miss him very much.

Acknowledgment

This tribute was originally published in *Caxtonian* (Journal of the Caxton Club of Chicago) 18:7 (2005), 1–3.

Conversations with Jay

David Meyer

"Jay Marshall," he would say when answering the telephone.

"Herr Marshall," I would say.

"Herr Meyer," he would answer. How and when these greetings came about, I can no longer remember. But he knew about my German background and my father's art studies in Munich in the 1930s and my travels with my parents to Germany in the 1980s. Jay was always interested in other people's lives.

Our telephone conversations began in the 1960s. I met his wife, Frances, while frequenting her magic shop in downtown Chicago. She suggested that I meet Jay, because he, a professional magician, and I, a boy in grade school, were both interested in the history of magic. He was traveling a lot in those years, appearing in theatres and on television; but when he was at home in Oak Park, I might catch him by phoning and he was never too busy to talk to me—and, no doubt, anyone else who called, and there certainly had to be many others.

When I was a teenager and acquired one of Houdini's own scrapbooks, Jay wanted to see it. I invited him to my family's house in Hammond and a day was arranged when he would stop by on his way to Detroit. I set out the scrapbook that morning and waited in excited anticipation for Jay to call from the highway to get directions. The hours passed, then the morning and, not having heard from him by mid-afternoon, I gave up in disappointment and left with my father to run errands. When we returned for supper I learned that Jay had called. He had left the magic shop later than expected: instead of ten o'clock he had left at four. We did not meet that day, and I was to learn that leaving late—for engagements and destinations—was routine for Jay.

Sometime in the 1980s, after his life in Chicago superseded life on the road, we talked on the phone more often. I was working in a family business and Jay would call me nearly every afternoon from the magic shop. By then he and Frances had relocated their home and shop to North Lincoln Avenue. Jay was busy appearing on local TV shows, publishing magic books and spending time with magicians and comedians who worked the Midwest circuits and came to Chicago to see him. In those years he complained about getting old (he was in his sixties) and we talked about our investments in the stock market.

During phone calls in the 1990s he began reminiscing about times and friends from earlier days. After each conversation I would rush to my little Tandy computer and attempt to record the stories as he told them. And when we spent time together, I often noted what had happened and what Jay had said. Jay, a wit and raconteur, was more of a Dr. Johnson than I could hope to be a James Boswell, but following are a few examples of what I saved:

26 September 1992. Jay phoned. Last night friends of puppeteer John Shirley gave him a seventy-fifth birthday party. John was told the occasion was a performance he was booked to do, so he was completely taken by surprise.

John's second (of three?) wives, Bonnie, was there and Jay told the story of driving down to Miami for a magic convention with the Shirleys years before. John, Bonnie and Jay made the fifteen-hundred-mile trip in twenty-four hours, each driving an hour at a time. On the trip Jay learned that Bonnie could neither read nor write due to dyslexia.

"I later bought some flash cards," he said, "and taught her how to read. As soon as she could spell 'divorce,' she got one."

Jay said that John Shirley remarked at the party that if he had one more wife and lived two more years, he could celebrate his golden wedding anniversary.

Figure 3. Jay and his famous puppet "Lefty" in 1951. The puppet now resides in the Smithsonian's Museum of American History.

19 October 1992. Jay recalled how he once gave a show in a West Virginia town when he was seventeen. He brought a hillbilly on stage who was twenty-four. "I put a sponge rabbit in the palm of my hand and put another in the palm of his hand. I closed my hand and [made] the rabbit disappear. I had the fellow open his hand and he held the two rabbits—and he punched me in the face. That's the last time I did that trick with a hillbilly."

15 August 1993. I had lunch with Jay and Fran. She ordered a cup of onion soup and kept lifting a spoonful to Jay's mouth for him to take. He went along with this several times, then told her he had had enough—no more. A few minutes later she pushed another spoonful at him, which he even more reluctantly accepted. Finally, he cursed. Yet she still persisted until he snapped, "*Please* stop doing that!" I said to him, "I think that's the first time in all the years I've known you that you've used the word 'please'."

"I used it for emphasis," he said.

[?] October 1993.   On a foray into Chicago to trade books with Jay he advised me that Fran was hosting a gathering of church counselors in the little theatre behind the magic shop. Bob Brown, a magician and reformed alcoholic, was making a pot of sloppy joes and beans for the group. Jay and I were invited to join them after our book dealing but I told Jay I wasn't interested in this kind of food and he said we needed to stop in to say hello anyway, and tell Fran where we were going. We entered the room just as the group finished singing together. Among them, Jay said, were priests, nuns and reformed drunks. When he had their attention, Jay said that we were declining their invitation because we were going out for a drink.

30 June 1994.   Mary Parrish told me that she had dinner with Jay and Fran the night before last and during the meal Fran said, "Oh, Jay, there's something about you which I just can't live without!"

18 July 1994.   Jay was describing, over the phone, the last night's show of the Society of American Magician's convention held in Chicago, which my wife Anita and I missed. He told me he had added a few lines to his Lefty act because the audience had seen him perform it so many times before. "You don't know how hard it is to put in a new line in an act you've been doing for forty years!" Apparently one of the changes was the addition of two words to the end of the first line Lefty sang: "If I had my way, dear, you'd never grow old—*Too late!*" "They gave me a standing ovation," Jay said, "and Tony Hacina took a photo of it. He's going to send me a print and I'll see who didn't stand."

6 September 1994.   Jay has always teased Anita about cooking chicken for dinner. He calls her "the chicken lady." When he phoned today I mentioned that Anita was making catfish for dinner. Jay assumed a deep southern drawl and said, "Catfish! Why that's *southern* chicken!"

Figure 4. In 1950, Jay appeared in "Great to Be Alive" on Broadway.

20 September 1994. At the Caxton Club meeting Jay was dressed in a blue blazer, starched white shirt, tie, and trousers. I told him he looked spiffy, better than I had seen him dress in a long time. He told me he was wearing his late friend Tommy Edwards' blazer from The Jockey Club. "I fit into all my friend's clothes," he said, almost sheepishly, and mentioned Tommy and John Shirley, both longtime friends who had recently died. After Bob Parrish died, Jay wore Bob's hat and scarf.

10 October 1994. Jay phoned to tell a joke, but I did not understand the punch line. "Tell your wife," he said. "She'll understand it. I should have told her in the first place."

Homage to a Pied Puzzler

He went on to say that he had enjoyed an excellent talk on the poetry of Robert Frost at a recent luncheon meeting of the Caxton Club. He explained that his knowledge of poetry did not extend beyond the line "There once was a man from Nantucket. . . ."

14 November 1994. Jay phoned Anita to ask her if she had ever seen a kitchen gadget that spun lettuce within a bowl to throw the water off after washing. He had found one in the Swedish Covenant Hospital thrift shop for $1.50 and had bought it. Anita told him that she had one and used it often. Jay's reaction: "Why does everyone else know about this except me?" He was not in a good mood. When Anita asked him if he wanted to talk to me, Jay said, "Is that really necessary?"

5 December 1994. Jay flew to the Showman's League convention in Las Vegas on a 4:30 A.M. flight because the person he was flying with weighed nearly 400 pounds and the man wanted to be assured of having an empty seat between the two of them.

10 December 1994. Jay tells me, "If air conditioning had been invented in stagecoach days, I would have preferred living then."

31 December 1994. Jay phoned and told Anita that he had come full circle. He was doing a show that night in exchange for dinner for himself and Fran. That was how he had started in show business.

18 January 1995. After telling Jay that I was fed up with a mutual friend's negativity toward me, Jay said, "Write him off. I'm thinking of writing you off myself."

16 March 1995. Jay is a member of the Society of the Fifth Line, a club whose members write limericks. He called and left on our answering machine his latest composition:

A young man with passions quite gingery
Tore a hole in his sister's best lingery
He pinched her behind
As he made up his mind
To add incest to insult to injury.

23 November 1995. Jay tells me that after a Thanksgiving party, as he was taking Fran back to her nursing home, she said, "What are you doing this for? You're going to end up in divorce court."

Early December 1995. Jay is in the hospital with a blood clot. I tell him that a friend of ours advises him to go to Mayo Clinic in Rochester if he needs surgery. The next day Jay tells me he's exhausted from phone calls and friends visiting him. "I should be in Rochester," he says. "No one knows me there."

4 January 1996. While driving from the magic shop to a restaurant for dinner, Jay told Anita and me of a woman he knew named Gloria. He couldn't recall her last name. She became pregnant, married the man, had two children, got a divorce, lived with her mother and played a piano in a cocktail lounge. One day, a Monday, she arrived late for work because she became ill on the bus. Jay quipped: "Sick in transit on Gloria's Monday."

21 May 1997. Jay called; he was frustrated because he could not find any tenor banjo strings. He had gone to the Carl Fischer Music Store in the Loop and wrangled with a young clerk, trying to explain what he needed, but without luck. He next asked for banjo picks, which they also did not have. He said he used to make them out of celluloid collar stays. "Do you have a pitch pipe for a banjo?" he asked the girl. "Not for a tenor banjo," she said. He complained to me about being obsolete.

In the last few years, starting about 2003, Jay phoned me every weekday to give me a stock market report on several companies we had both invested in. He did not want to talk about other subjects; he gave me the Dow Jones Industrial average for the day, the closing price on six or seven stocks and the conversation, as far as he was concerned, was over. When Anita and I were not home to receive a call, he left a message. Anita's recorded voice on our answering machine instructed a caller to "press star five one if you want to leave a FAX." Jay took this information as his identity and when he called he would begin his report by saying, "This is Star Five One."

One of the last times he called I made the mistake of telling him that Anita was baking oatmeal chocolate chip cookies. Jay liked them but I had to tell him that I wasn't coming up that week to have lunch with him so he wouldn't be getting any.

Figure 5. Jay in a publicity photo, probably from the 1970s.

A few days later, just before Easter, Anita answered the phone and received Jay's stock report, as she often did when I was not available.

"Happy Easter," she told him.

"Happy oatmeal cookies," he replied.

In all the decades I knew Jay Marshall, his signature sign-off at the end of a phone conversation was "Keep in touch."

I wish I could.

Acknowledgment

This tribute was originally published in *Caxtonian* (Journal of the Caxton Club of Chicago), 18:7 (2005), 4–6.

Part I

History and Hoaxes

Sam Loyd's Most Successful Hoax

Jerry Slocum

Martin Gardner called Sam Loyd "America's Greatest Puzzlist." And Loyd is famous for the numerous wonderful puzzles that he invented, such as the Trick Mules (Figure 1, left), Get Off the Earth Puzzle Mystery (Figure 1, center), and the thousands of delightful puzzles included in his *Cyclopedia* (Figure 1, right) as well as the engaging stories he used to pose his puzzles.

However, Loyd also has a reputation for taking credit for puzzles created by others. Henry Dudeney complained numerous times that Loyd did not give him credit for puzzles that Dudeney had created.

And sometimes his delightful stories accompanying his puzzles were beyond a mere exaggeration, they were a hoax, completely false. Loyd's book about the Tangram puzzle, called *The 8th Book of Tan* (shown in Figure 2), included a extensive, but bogus, history of the puzzle that claimed that it was 4,000 years old. In fact, the puzzle is actually about 200 years old.

Although Sir James Murray, Editor of the *Oxford English Dictionary*, exposed Loyd's false Tangram history in 1910, seven years

Figure 1. Two great puzzles and the *Cyclopedia of 5000 Puzzles* by Sam Loyd.

Figure 2. *The 8th Book of Tan* by Sam Loyd

Figure 3. The Pigs-in-Clover puzzle. (See Color Plate I.)

Figure 4. Judge's political Pigs-in-Clover cartoon. (See Color Plate II.)

Charles M. Crandall holding his best-
selling puzzle, "Pigs in Clover." A
model made from his building blocks is
behind the puzzle. (1889)

Figure 5. Charles Crandall, inventor of Pigs-in-Clover.

after the book was published, Loyd's hoax is still occasionally reported in publications and websites.

An interview with Sam Loyd in the *Lima* (Ohio) *Daily Times* on
January 13, 1891, provided Loyd an opportunity to plug a new
puzzle of his, named "Blind Luck." Loyd also mentioned in the
interview, for the first time, that he was the inventor of Pigs-in-
Clover (shown in Figure 3), which had been a puzzle craze in 1889,
just two years earlier. He also claimed that he invented the popular
game Parcheesi, and the 14-15 Puzzle.

Let's look at each of Loyd's claims of invention.

From mid-February to May 1889, Pigs-in-Clover was a huge
puzzle craze in the United States. Three weeks after it came out,
the *Waverly Free Press* reported that the Waverly "toy works were
turning out 8,000 a day and they are 20 days behind in their orders." The puzzle reached the US Senate by mid-March and be-

OATH.

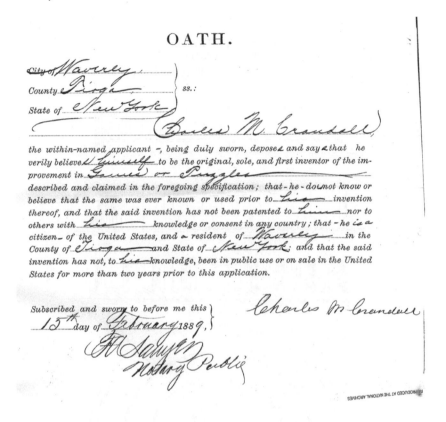

Figure 6. Crandall's oath of invention.

came a metaphor for politics, as we see in Figure 4. Mark Twain mentioned it in his book *The American Claimant*.

The question of who invented the puzzle was, at that time, not in dispute. Charles Crandall (see Figure 5) was reported to be the inventor in numerous newspaper articles in 1889. And Sam Loyd's name was never found in connection with the puzzle at the time of the craze.

We checked further and found that Crandall applied for a patent in February 1889 and he signed an oath shown in Figure 6 that he was the sole inventor.

The patent for the puzzle, shown in Figure 7, was issued to Crandall on September 10, 1889. It is surprising that Loyd had

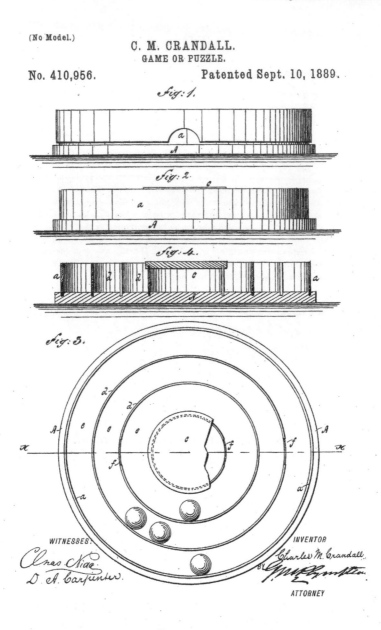

Figure 7. Crandall's patent for Pigs-in-Clover.

Figure 8. Hindu Parchessi players, eighteenth century.

the audacity to claim that he invented Pigs-in-Clover less than two years after the puzzle craze ended and with all the newspaper reports that Crandall had invented it.

And it is even more surprising that he succeeded in making the world believe his claim.

We also checked on Loyd's claim that he invented the game Parcheesi. The *Chicago Tribune* reported in May 1893 about an exhibit on the history and evolution of games at the World Columbian Exposition being held in Chicago. One of the games exhibited was Parcheesi (see Figure 8), and the game was described as coming from India where it had been played since the fourth century. The game company Selchow & Righter bought the rights to Parcheesi in 1870 and obtained a trademark for it in 1874. There was no evidence of any involvement by Sam Loyd.

Now let's examine Loyd's claim that he invented the 15 Puzzle. According to newspaper reports, the first commercial 15 Puzzle was made by Matthias J. Rice in Boston and first sold in Boston and Hartford Connecticut two weeks before Christmas 1879. It was called "The Gem Puzzle."

Notice in Figure 9 that the 15 blocks are not constrained in the box and the instructions are very simple.

Place the blocks in the box irregularly, then move until in regular order.

Rice said in an interview that once he succeeded in having a store in Boston sell the puzzle, he could never keep up with orders!

Figure 9. Matthias Rice's Gem Puzzle was the first commercial 15 Puzzle.

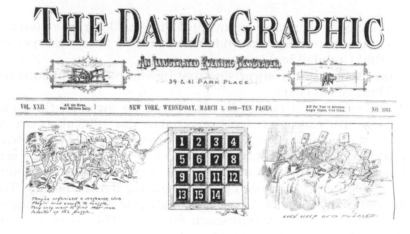

Figure 10. *The Daily Graphic* at the peak of the 15 Puzzle craze, March 3, 1880.

By mid-February 1880, sales took off from Boston and New York and by mid-March the entire country was involved in an enormous 15 Puzzle craze.

From the beginning, the craze was initiated and fueled by the phenomenon that one time when you tried, you could solve it—and the next time it seemed impossible. The big problem was when you were able to get all the numbers correctly placed except the numbers 15 and 14, which were reversed—the only ones not in

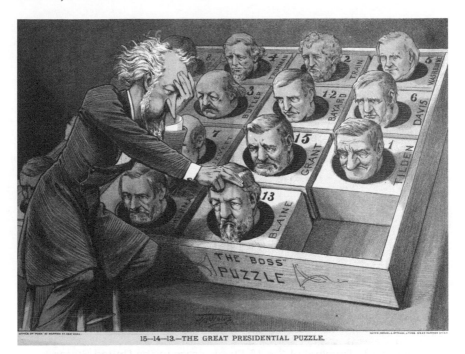

15—14—13.—THE GREAT PRESIDENTIAL PUZZLE.

Figure 11. *PUCK* magazine cartoon about the 1880 presidential election. (See Color Plate III.)

"regular order." Still, there were many people that claimed it was always solvable. During March the puzzle craze generated songs, theatre productions, and political cartoons, such as the ones in Figure 10.

America's dominant political satire magazine, *PUCK*, published a full-page color political cartoon about the selection of candidates for the 1880 Presidential Election featuring the 15 Puzzle. The cartoon is shown in Figure 11.

To get a feeling for how the 15 Puzzle craze started and spread in the US, let's look at the timeline below showing the start of 15 Puzzle sales and how the puzzle craze that it created rapidly spread and enveloped the whole country.

As you can see in Figure 12, the craze spread from New York to San Francisco in about three weeks and ended in most of the United States by mid-April, 1880; but an equally big 15 Puzzle craze in Europe had just begun.

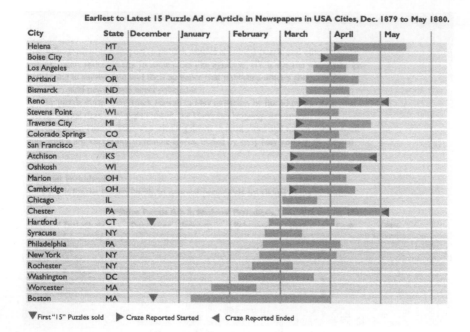

Earliest to Latest 15 Puzzle Ad or Article in Newspapers in USA Cities, Dec. 1879 to May 1880.

▼ First "15" Puzzles sold ▶ Craze Reported Started ◀ Craze Reported Ended

Figure 12. Timeline of the 15 Puzzle craze in the United States.

Now let's look at Sam Loyd's involvement in the 15 Puzzle. As described earlier, Sam Loyd first mentioned that he invented the 15 Puzzle in an interview published in January 1891, *more than ten years after the 15 Puzzle craze had ended!* He continued to make the same claim for the next 20 years, before his death in 1911, in at least 23 articles, ads, and interviews that we found.

We looked for but did not find any articles, interviews, or publications that mentioned any Loyd involvement with the 15 Puzzle during the 1870s or 1880s. The list in Table 1 shows a sample of the articles and interviews that credited Loyd with the invention. We will discuss two items, the seventh and the last in the list, in detail a bit later.

After Loyd's death in 1911, nine of the eleven obituaries that we found mentioned his invention of the 15 Puzzle, including all of the ones shown in Table 2, and eight of eleven mentioned that he invented Pigs-in-Clover as well.

Date	Publication	Title
January 13, 1891	*Lima Daily Times*	"He Invents Puzzles"
February 1891	*Manufacturer & Builder* Magazine	Note about "Blind Luck"
1892	Ad for Piper Heisdick Tobacco	"Our '93 Challenge"
May 11, 1893	*Indiana Country Gazette*	Charity Fund request
October 14, 1893	*Tit Bits*	Sam Loyd Puzzle
July 1894	*Our Illustrated Press*	Loyd's "Our Puzzle Corner"
January 4, 1896	*Illustrated American*	"The Famous 15 Block Puzzle"
March 22, 1896	*Brooklyn Daily Eagle*	Loyd's first puzzle column
May 31, 1896	*New York Sun*	Loyd interview
June 24, 1896	Sam Loyd Letterhead	Loyd letter to Mr. Raynor

Table 1. Articles and interviews that credit Loyd for the invention of the 15 Puzzle.

Date	Publication	Title
April 12, 1911	*New York Times*	"Sam Loyd, Puzzle Man, Dies"
April 12, 1911	*Chicago Tribune*	"Puzzle King Sam Loyd Dies"
April 12, 1911	*Mansfield News*	"Noted Puzzler Dies"
April 22, 1911	*Scientific American*	"Samuel Loyd"
May 1911	*The American Magazine*	"Sam Loyd"
June 7, 1911	*Gettysburg Compiler*	"The Puzzle King is Dead"
June 23, 1911	*Stevens Point Journal*	"The 'Pigs in Clover' Man"

Table 2. Obituaries of Sam Loyd.

Date	Publication
December 1926	*Nineteenth Century*
January 1930	*Science & Invention*
February 24, 1934	*New York Herald*
September 13, 1941	*Christian Science Monitor*
March 5, 1943	*Hammond Times*
August 1957	*Scientific American*
January 1984	*Science Digest*

Table 3. Articles that credit Loyd with the 15 Puzzle invention.

Year	Author	Title
1911	Ahrens	*Mathematical Spiele*
1914	Loyd	*Cyclopedia of Puzzles*
1931	Dudeney	*Puzzles and Curious Problems*
1943	Malone (ed.)	*Dictionary of American Biography*
1978	van Delft	*Creative Puzzles of the World*
1995	—	*Merriam-Webster's Biographical Dictionary*
1996	Pickard	*The Puzzle King*
2005	—	*Encyclopedia Britannica*

Table 4. Books that credit Loyd with the 15 Puzzle invention.

SAM LOYD,

Journalist and Advertising Expert,

ORIGINAL
Games, Novelties, Supplements, Souvenirs,
Etc., for Newspapers.

Unique Sketches, Novelties, Puzzles, &c.,
FOR ADVERTISING PURPOSES.

Author of the famous
" Get Off The Earth Mystery," " Trick Donkeys,"
"15 Block Puzzle," " Pigs in Clover,"
" Parcheesi," Etc., Etc.

P. O. BOX 1821.

New York, *June 24* 1896

Figure 13. Sam Loyd's letterhead, June 1896.

And after his death many articles, books, biographies, and encyclopedias credited Loyd for the 15 Puzzle invention, including the 2005 Encyclopedia Britannica. Some of these are listed in Tables 3 and 4.

Sam Loyd's own letterhead—at least from 1896 until 1903—claimed that Loyd was the "author" of the "15 Block Puzzle," "Pigs in Clover," and the game Parcheesi! Loyd's letterhead as of June 1896 is shown in Figure 13

Now let's look at the first article by Sam Loyd where he challenged readers to solve the 15 Puzzle. Sam Loyd's article in the *Illustrated American* on January 4, 1896 included a sketch of a puzzle solver drawn by Loyd (see Figure 14), and it was the first

The Famous 15 Block Puzzle.

WHETHER thousands of persons really went insane or not over the "15 Block Puzzle," which had such a phenomenal run some twenty years ago is difficult to prove, but it is a fact that the entire world—from the highest to the lowest, wise as well as foolish—went mad over the tantalizing mystery of the fifteen little blocks.

Many remarkable stories are told of prominent people who were held in thraldom by the fascinating mystery. One clergyman of note purchased the puzzle on Park Row in the morning, and stood on the sidewalk moving the numbers around until midnight. An editor of a leading daily was discovered in a restaurant, where in a fit of abstraction he had cut his pie into little pieces, and was pushing them about on the plate. One mild form of insanity which did develop to an alarming degree pertained to the history of the puzzle and to the veracity of such as mastered it. More

pasted on to small pieces of cardboard, so as to make a very convenient and durable puzzle game. Of course it is understood that the movable blocks are placed on the board with the 15, 14 standing transposed as shown.

The object of the problem is to move the blocks one at a time, and bring about an absolutely perfect sequence of figures, with the 14 and 15, as well as all other numbers, in their regular order. As a number can only be moved to the one vacant square, it becomes a very simple matter to record the required play by writing out the blocks played, as, for example, 14, 11, 10, 15, 13, etc., etc., to the desired arrangement.

It requires but a few minutes' experience, in an honest endeavor to untangle the transposition of the 15 and 14, to give one a taste of the fascination that made this remarkable puzzle so popular as to be referred to in parliamentary debates, or leading editorials throughout the civilized world.

Patience and chance are more potent than science or mathematics in the solution of this curious problem, of which it was well said "the more you study the less you know." You straighten out one little transposition only to find that an entanglement has been pro-

Figure 14. Sam Loyd's first article where he challenged readers to solve the 15-14 Puzzle and offered a $1,000 reward, to be divided among the solvers.

The mystery of the 15 Block Puzzle was at once developed and became a craze. I give it as originally promulgated in 1872.

As every one claimed to have discovered the mystery, and the author is desirous of learning if any, or how many, kept a record of the sequence of moves which effects the solution, he offers to divide a thousand dollars equally among all who send correct answers within the next thirty days.

The board is reproduced just as originally drawn, with the numbers marked on the fifteen squares, showing how the blocks must be arranged at the commencement. The blocks may be cut apart, and

Figure 15. Loyd claims he posed the 15 Puzzle in 1872.

time Loyd challenged his readers to solve the 15-14 Puzzle. He offered a $1,000 reward for the solution!

Loyd wrote (see Figure 15), "I give it as originally promulgated in 1872. ... The board is reproduced just as originally drawn, with the numbers marked on the fifteen squares, showing how the blocks must be arranged at the *commencement*" (author's italics). The block positions at the start are illustrated in Figure 16.

And a reward of $1,000 was offered to be divided "equally among all who send correct answers."

Figure 16. Loyd's definition of the block positions at the start, with reversed positions for the 15 and 14 blocks.

Figure 17. Solution to Loyd's stated problem.

Loyd's instructions are: "The object of the problem is to move the blocks one at a time, and bring about an absolutely perfect sequence of figures, with the 14 and 15, as well as all other numbers, in regular order."

Considering that Loyd was a professional puzzle designer who had a lot of experience challenging readers, it is obvious that he had never proposed this problem to readers in 1872 or at any other time because the problem *as he posed it* was indeed solvable!

The solution in Figure 17 was first mentioned in an article in the *Worcester Evening Gazette* of January 29, 1880, and the

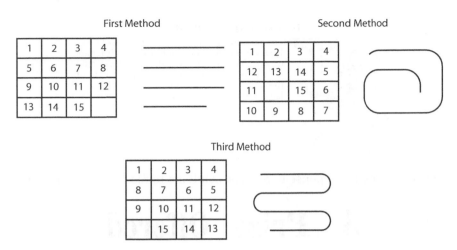

Figure 18. Persifor Frazer presented 48 solutions of the 15 Puzzle to the American Philosophical Society on March 19, 1880.

same solution by MAME and J. D. Warren was published in the *Rochester Democrat and Register* on February 23, 1880 (16 years before Loyd's article). W. W. Rouse Ball published this solution in his book *Mathematical Recreations and Problems* in 1892.

And Persifor Frazer presented 48 solutions for the 15 Puzzle to the American Philosophical Society on March 19, 1880. Part of Frazer's work is shown in Figure 18.

Of course all of these solutions comply with the instruction that the blocks must be placed in "regular order"—but not the "regular order" that Loyd had in mind.

The key aspect of the problem that Loyd apparently was not aware of was that the location of the empty space must be specified when posing the 15 Puzzle!

So what did Sam Loyd discover about his 15 Puzzle and his $1,000 reward for a solution?

He was a bit surprised! As shown in Figure 19, Loyd wrote: "it is safe to say that the thousand-dollar prize would have to be divided into the smallest coin of the realm to be portioned among all who have worked out solutions of some sort."

But did he pay the $1,000 reward for the solutions? Of course not! He wrote, "Many succeeded in getting the correct position, but failed to record the sequence of plays."

The Famous 15 Block Puzzle.

THAT tantalizing mystery of the 15 blocks has been causing no end of interest, and it is safe to say that the thousand-dollar prize would have to be divided into the smallest coin of the realm to be portioned among all who have worked out solutions of some sort.

Figure 19. Loyd's response to readers' solutions to his 15 Puzzle problem.

A Prize Offered.

—

THE "15" PUZZLE.

—

WHO CAN DO IT?

—

I Offer a Set of $25.00 Teeth,

"THE BEST,"

on Rubber or Celluloid, and made by my NEW IM-PRESSION, warranted and perfectly adjusted,

AND $100 CASH,

To the Successful Competitor.

☞ Open to the whole world,

Set the numbers in regular order, from 1 to 15, then transpose Nos. **14** and **15** and proceed. When finished the blank must be after the No. 15.
This offer stands good for one month.

DR. CHAS. K. PEVEY,

Pevey's Dental Rooms,

WORCESTER, MASS.

Figure 20. The first reward for solving the 15 Puzzle was a set of false teeth and $100 offered by Dr. Pevey in January 1880.

Was he the first to offer a financial reward for a solution to the 15 Puzzle? Far from it! The first financial reward for a solution to the 15 puzzle was offered by Dr. Pevey, a dentist in Worcester, Massachusetts, on January 20, 1880, as shown in Figure 20.

Notice how Pevey defined the finished state: "When finished the blank must be after the No. 15." Thus, he avoided Loyd's error. Three weeks later he upped his reward to $1,000 for the solution, beating Loyd to the $1,000 reward by about 16 years.

Why did Dr. Pevey offer rewards for the solution to the 15 Puzzle? Dr. Pevey explains in the *Worcester Evening Gazette*, January 31, 1880:

> The reason why I wanted you all to help me work out the puzzle was to convince that girl. You see she said she worked it out; she knew she did, and if I said she did not, I simply doubted her veracity. Now to doubt the word of a young lady is high treason and of course should be punished as such, so I stopped to think how I could convince her (without putting it into words) that she did not do what she said she did. From her looks I made up my mind it was no easy task and would probably require the whole population of Worcester to help me. For she knew she did it! And that you know would ordinarily settle it, but take what it would or cost what it would, she must be convinced, but now that we have all given the matter a weeks careful study, and without a single favorable result. Probably she will no longer contend that she did it.

> At first some of us, as you know, rather held to it that it could be done, and that perhaps she was right. But now that we are all of one mind, that it can not be done, and that we were mistaken, we will laugh over our week's fun and proceed to business again.

> Respectfully,
> Chas. K. Pevey

Pevey's Dental Rooms, cor. Main and Pleasant Streets, Worcester, Mass.

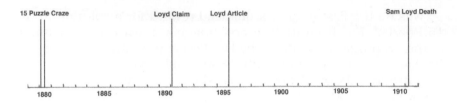

Figure 21. Timeline of key dates in Sam Loyd's hoax.

The timeline in Figure 21 summarizes the key dates in Loyd's hoax:

- 1880: 15 Puzzle craze worldwide ended by July 1880,

- 1891: Loyd's first claim that he invented the 15 Puzzle,

- 1896: Loyd's first article challenging readers to solve the 15 Puzzle,

- 1911: Loyd's death.

Here are our conclusions about Sam Loyd's 15 Puzzle hoax:

- Sam Loyd did not invent the 15 Puzzle.

- He had no role in the 15 Puzzle craze.

- Loyd's campaign to take credit for the invention began in 1891—ten years after the 15 Puzzle craze ended.

- At least 23 articles and interviews repeat Loyd's claim of invention of the 15 Puzzle.

- Loyd's campaign was very successful.

- After his death in 1911, obituaries, books, articles, websites, and even encyclopedias 94 years later falsely credit him with inventing the 15 Puzzle.

Since Martin Gardner has written that Sam Loyd invented the 15 Puzzle, I thought it would be interesting to get his reaction to our findings, so I sent a galley proof of *The 15 Puzzle* to him.

Martin Gardner commented:

> Did Sam Loyd, America's greatest puzzle maker, invent the notorious 14-15 sliding block puzzle? He claimed he did but the claim was a total lie. Loyd had nothing whatsoever to do with either the puzzle or its popularity.

Acknowledgment

This article is based on findings reported in *The 15 Puzzle*, by Jerry Slocum and Dic Sonneveld, published by The Slocum Puzzles Foundation in 2006, and was first presented in lectures at the Seventh Gathering for Gardner, March 2006; the Convention of the Association of Game and Puzzle Collectors, March 2006; and the Twenty-Sixth International Puzzle Party, July 2006.

Part II

Numbers and Shapes

The Seven-Colored Torus: Mathematically Interesting and Nontrivial to Construct

sarah-marie belcastro and Carolyn Yackel

Any map drawn on the torus can be colored in seven or fewer colors. There is a map on the torus that requires seven colors. We explain why the number seven arises, and describe our efforts to design knitted and crocheted tori exhibiting this map.

Why the Seven-Colored Torus is Cool

The well-known Four Color Theorem, conjectured in 1852[1] but proven only in 1976 [1,2], states that given any map on the sphere composed of contiguous countries, a cartographer needs only four colors to color the map so that any two countries that share a border are colored differently. The Empire Problem is an extension of

---

[1]According to most historical accounts, Francis Guthrie posed the Four Color Problem in 1852 and transmitted it to his brother Frederick who in turn asked Augustus DeMorgan about it, who in turn asked William Rowan Hamilton about it....

Figure 1. A torus in 3-space (left), and a torus drawn in 2-space (right).

this problem in which each country has m connected components. Heawood showed [4] that $6m$ colors suffice and his conjecture that they are necessary was proven 94 years later [5]. A different extension, known as the Earth-Moon Problem [6], allows each country to have two connected components (one on the earth and one on the moon). By Heawood's Empire result, 12 colors suffice. In 1980, Gardner reported an example requiring nine colors[2] [3]. However, the question remains open as to whether or not more colors are necessary. Yet another extension, and the one that we will examine here, asks how many colors are required for a map drawn on a nonspherical planet.

The simplest (orientable) non-spherical surface is the torus, shown in Figure 1 in its doughnut-like form and as a rectangle with opposite sides identified (which is much easier for drawing maps). There are also surfaces with many doughnut holes.

The problems mentioned above are studied by reformulating them as problems in graph theory, where each country is replaced by a vertex and two vertices are connected by an edge exactly when the countries they represent share a nontrivial boundary. Coloring a map is then equivalent to vertex-coloring an associated graph. We call the map and the graph dual to one another. Further, a graph drawn on a surface without edges crossing is said to be embedded in the surface.

In order to examine the problem of vertex-coloring graphs on nonspherical surfaces, we consider the complete graph on k vertices (denoted K_k), which has every pair of vertices connected by an edge. Every graph is contained in some complete graph, so K_k gives us the worst-case scenario for coloring graphs. Thus, the question is really, what is the largest k for which K_k embeds on a surface

---

[2]The example cited by Gardner is Thomas Sulanke's from 1974 and is otherwise unpublished.

with n holes? An upper bound is given by the following inequality:

$$k \leq \left\lfloor \frac{7 + \sqrt{1 + 48n}}{2} \right\rfloor .$$

This is called the Heawood bound [4]. Looked at another way, given the the number of holes in a surface, n, this bound gives the greatest number of colors needed to color any embedded graph on that surface or the dual map to the graph. Notice that if S is a torus, where $n = 1$, the Heawood bound shows that $k \leq 7$. Indeed, K_7 embeds in the torus, as we will exhibit in the next section. Thus, we know that not only does there exist a map that requires seven colors, but that is the greatest possible number of colors needed for a map on the torus. The bound is also sharp for all surfaces except the Möbius band[3] [7].

Design Plans for Seven-Colored Tori

In celebration of G4G7's theme of "seven," we set out to construct a pair of tori, one of which would be knitted and the other crocheted, and one of which would display K_7 and the other its dual map. The mathematics behind the engineering involved in using the two fabric-making techniques motivated the project. Below we describe successive phases in our modeling of the problem, which ultimately led to workable patterns.

The visual goals were to exhibit a seven-colored torus and a torus with K_7 embedded so that each torus individually was a clear representation of its mathematical pattern and so that together the pair clearly showed duality. Therefore, some design components, often disregarded in representations of the seven-colored torus, became essential to our renderings. For example, the vertices of the K_7 torus needed to be easily visible, rather than placed along the inner surface of the torus, where they would be hard to see. In addition, we required that they be symmetrically placed and thus aesthetically pleasing. Because seven is a prime number, there is no symmetric arrangement of the vertices that allows for offsetting the vertices from each other. Therefore, all seven vertices must be arranged along the outermost surface of the torus; such an embedding is shown in Figure 2 (left).

---

[3]This result was completed, primarily by Ringel and J. W. T. Youngs, in 1968; Youngs died in 1970 and Ringel published the work in 1974.

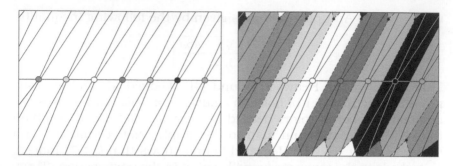

Figure 2. An embedding of K_7 on the torus (left), and an embedding of K_7 on the torus along with its dual map (right). In our embedding diagrams, the middle corresponds to the outer region of the torus, while the top and bottom of the diagram are identified and correspond to the inner surface of the torus. (The left and right sides are also identified.)

A similar visibility issue arises in creating the dual map, which is initially done by placing one vertex in each face of the original graph embedding, and drawing edges between these new vertices in one-to-one correspondence with the edges of the original graph, as shown in Figure 2 (right). The careful torus inspector will want to ensure that each color shares a boundary with each other color; hence, the best map will have none of these boundaries entirely hidden on the inner surfaces of the torus. Still, it is aesthetically desirable for the outer surface of the torus to show only a set of stripes. The solution to this problem is the transformation shown in Figure 3. Notice the shift in edge lengths of the countries of the map so that the boundaries lie centrally along the top and bottom surfaces of the torus.

The diagrams in Figures 2 and 3 do not reflect the actual shape of the torus—in reality, the innermost surface of the torus will be smaller than the outermost surface, and those are shown as the same size. In order to reflect the curvature of the tori in our diagrams, we needed to think about the practical knitting and crocheting aspects of the project.

We had decided that Carolyn would crochet the seven-colored map, and sarah-marie would knit the embedding of K_7, because knitting edges on the graph is technically easy in terms of switching yarns, whereas switching colors frequently in crochet is diffi-

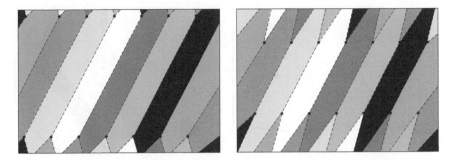

Figure 3. A dual map to K_7 on the torus (left), and a seven-colored map on the torus (right). (See Color Plate V.)

cult. The knitted curvature was created using the common technique of short rows. Though this technique is not common in crochet, the obvious solution was to mimic this design. In an attempt to do so, two problems arose. First, short rows involve turning the work. Knitters generally then switch between rows of knit stitching and rows of purl stitching. These paired stitches look like each other when reversed, so that the back side of a purled row looks like the front side of a knitted row. Crochet has no such matched pairs of stitches; therefore, the solution to this problem was to switch between right- and left-handed crochet. Left-handed crochet does not require reversing the work and looks essentially like right-handed crochet. The second problem faced in both knit and crochet was that the starting and ending locations of the short rows needed to be offset from one another in order to avoid rows of gaps in the fabric, in the case of knitting, or giant chasms surrounded by bumpy singularities, in the case of crochet. The difference in severity of the problem arose because of the larger height of a single crochet stitch. Ultimately, this gives the knitted pattern the appearance of two flat sections of different lengths (the outer and inner surfaces of the torus) joined by narrow bridges on the top and bottom where the short rows' edges lined up, as in Figure 4 (left). The crocheted pattern has a broader bridge section, corresponding to larger offsets, as in Figure 4 (right).

Having now determined the basic shape of the patterns, we determined suitable dimensions and made the grids for each pattern. Miraculously, the knitted and crocheted gauges (grid proportions) were the same. Determining the patterns themselves was surpris-

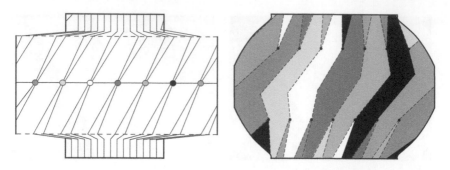

Figure 4. Embeddings of K_7 (left) and its dual map (right) that reflect curvature.

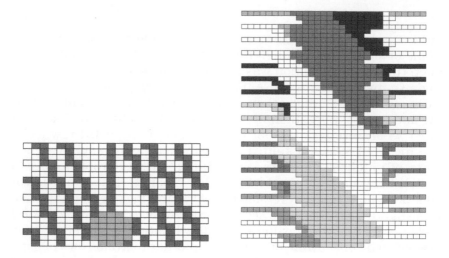

Figure 5. The knitting pattern for 1/7 of the torus (left), and the crochet pattern for 3/7 of the torus (right). (See Color Plate VI.)

ingly difficult, primarily because of the need to discretize the continuous mathematical models we had so that each stitch could be knit or crocheted in a single color of yarn. Both patterns required that we calculate (or at least approximate) slopes; on the knitted torus, the six edges emanating from each vertex have three different slopes, and on the crocheted torus the six boundaries of each country have three different slopes. The slopes could not be calculated in any straightforward way because of the bridge stitches

Figure 6. The completed tori, positioned dually. (See Color Plate VII.)

and short rows. Additionally, the vertices on the knitted torus were of nonzero height and width, throwing off slope calculations further. Complicating matters for the crochet pattern are the facts that each country extends across 3/7 of the outer surface of the torus and that each hexagonal country is symmetric under a 180° rotation but the placement of the short rows is not similarly symmetric. The resulting patterns are shown in Figure 5, followed by a photograph of the completed tori in Figure 6.

Bibliography

[1] K. Appel and W. Haken. "Every Planar Map Is Four Colorable. Part I. Discharging." *Illinois J. Math.* 21 (1977), 429–490.

[2] K. Appel, W. Haken, and J. Koch. "Every Planar Map Is Four Colorable. Part II. Reducibility." *Illinois J. Math.* 21 (1977), 491–567.

[3] M. Gardner. "Mathematical Recreations: The Coloring of Unusual Maps Leads Into Uncharted Territory." *Scientific American* 242 (1980), 14–22.

[4] P. J. Heawood. "Map Colour Theorems." *Quart. J. Pure Appl. Math.* 24 (1890), 332–338.

[5] B. Jackson and G. Ringel. "Solution of Heawood's Empire Problem in the Plane." *J. Reine Angew. Math.* 347 (1984), 146–153.

[6] G. Ringel. *Färbungsprobleme auf Flachen und Graphen.* Berlin: Deutsche Verlag der Wissenschaften, 1959.

[7] G. Ringel. *Map Color Theorem,* Die Grundlehren der mathematischen Wissenschaften, Band 209. New York-Heidelberg: Springer-Verlag, 1974.

A Property of Complete Symbols: An Ongoing Saga Connecting Geometry and Number Theory

Peter Hilton, Jean Pedersen, and Byron Walden

This is a story that really begins in ancient Greece. Around about 350 B.C., the Greeks were fascinated with the idea of constructing regular N-gons with Euclidean tools (straight edge and compass). They were successful in constructing regular N-gons for $N = 2^c N_0$, with $N_0 = 1$, 3, 5, or 15. Of course, we need $N \geq 3$ for the polygon to exist. Naturally the Greeks would have liked to answer the question for every N, but, in fact, it seems that no further progress was made until about 2000 years later, when Gauss (1777–1855), the prince of mathematicians, completely settled the question by proving that a Euclidean construction of a regular N-gon is possible *if and only if* N is of the form

$$N = 2^c M,$$

33

where M is a product of *distinct* Fermat primes. A *Fermat prime* is a number, of the form $F_n = 2^{2^n} + 1$, that is prime.

Gauss' proof is remarkable in that it tells us when a Euclidean construction is possible, provided we know which Fermat numbers are prime—which we don't! However, we do know that the following Fermat numbers are prime:

$$F_0 = 3, \quad F_1 = 5, \quad F_2 = 17, \quad F_3 = 257, \quad F_4 = 65537.$$

The great Swiss mathematician Euler (1707–1783) showed that $F_5 = 2^{32} + 1$ is not prime and, although many composite Fermat numbers have been identified, to this day no other Fermat numbers have proven to be prime, beyond those listed above. Thus, even with Gauss' contribution, a Euclidean construction of a regular N-gon is known to exist for very few values of N; and, even for these N, we do not know, in all cases, an explicit construction.

Folding N-gons

In the middle of the twentieth century one of the authors (Pedersen) discovered a systematic folding procedure, using straight strips of paper (like adding machine tape), that produced some regular polygons to any desired degree of accuracy.[1] Shortly thereafter she began working with Hilton and they described a systematic method of producing a regular b-gon, to any desired degree of accuracy, for *any* number $b \geq 3$. The procedure will first be described for b odd and then extended to deal with the case of b even.

Thus, Pedersen and Hilton redefined the Greeks' original problem. They decided they would be content to be able to produce approximations to regular b-gons as long as they could depend on the error becoming smaller and smaller. This seems reasonable since Euclidean constructions are only perfect in the mind—after all, what is actually produced is a function of how sharp your pencil is, how steady you hold the compass, and how carefully you place the ruler. Thus, even with Euclidean constructions, there are inevitable inaccuracies, due to human error. As you will soon see, in the systematic folding procedures we use, every correct fold that

---

[1]The paper-folding result came about by trying to construct hexa-flexagons, written about by Martin Gardner in January of 1957. Gardner later wrote about Pedersen's braided polyhedra in his book *Wheels, Life, and Other Mathematical Amusements* (both articles may be found in [1]).

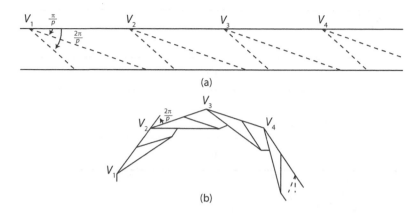

Figure 1. (a) Crease pattern for the strip of tape. (b) Folded strip forming a p-gon.

is made always cuts any previous error in half—and, as you would expect, this produces very respectable (even if not completely perfect!) regular polygons.

All this work constructing regular polygons leads to some fascinating number theory involving the *Euler totient function*.[2] To deal with some of the questions arising, Pedersen and Hilton brought in the third author, Walden. Now, back to the paper-folding.

First, we need to explain how a suitably creased strip of tape may be folded by what we call the *FAT-algorithm* to produce a regular convex p-gon. Figure 1(a) shows a strip of paper on which the dotted lines indicate certain special crease lines. Assume that the crease lines at the vertices labeled V_1, V_2, ..., which are on the top edge, form identical angles of $\frac{\pi}{p}$ (π radians $= 180°$) with the top edge with an identical angle of $\frac{\pi}{p}$ between the two downward crease lines. Now, if you fold first on the longer crease line coming from V_1 and then (twisting the paper in the same spiral direction) on the shorter line coming from V_1, you will see that the top edge of the strip has rotated through an angle of $\frac{2\pi}{p}$. Repeating the process at the points V_2, V_3, and V_4 will produce the portion of a

---

[2]The Euler totient function, Φ, counts the number of positive numbers less than a positive number b that are relatively prime to b. It is a well-known result of number theory that if p is prime then $\Phi(p^n) = (p-1)p^{n-1}$. Furthermore, for all mutually prime positive numbers k and ℓ, $\Phi(k \cdot \ell) = \Phi(k)\Phi(\ell)$. Thus we can calculate $\Phi(m)$ for any positive integer m.

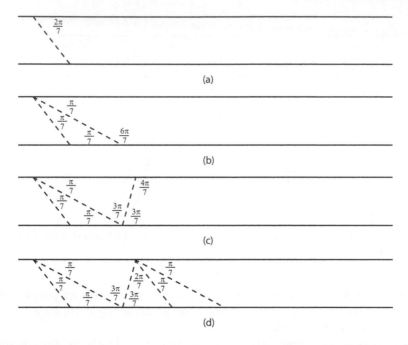

Figure 2. The optimistic strategy for $\frac{\pi}{7}$: (a) Creasing the initial angle $\frac{2\pi}{7}$. (b) After folding down to bisect $\frac{2\pi}{7}$. (c) After folding up to bisect $\frac{6\pi}{7}$. (d) Ending up with pairs of adjacent angles measuring $\frac{\pi}{7}$.

regular polygon shown in Figure 1(b). When this process of *folding and twisting*, which we call the FAT-algorithm, is repeated p times, the top edge will have rotated through an angle of 2π and thus, a regular p-gon will emerge.

Since the 7-gon is the polygon with the smallest number of sides for which no Euclidean construction exists (and also because 7 was the theme of the G4G7 conference!), we describe how to produce a strip of paper on which the smallest angle at the top is $\frac{\pi}{7}$. We are seemingly faced with great difficulty in creating the necessary crease lines along the top of the tape. However, let us invoke what we call our *optimistic strategy*. We assume that we *can* crease an angle of $\frac{2\pi}{7}$ (certainly we can come close), as shown in Figure 2(a). Assuming that we have this angle of $\frac{2\pi}{7}$, it is then easy to fold the top edge of the strip DOWN to bisect this angle, producing two adjacent angles of $\frac{\pi}{7}$ as shown in Figure 2(b). Then, since we are content with this arrangement, we go to the bottom of the tape where we observe that the angle to the right of the last crease line is $\frac{6\pi}{7}$—and we decide, as paper-folders, that we will always avoid leaving even multiples of π in the numerator of any angle next to an edge of the tape, so we bisect this angle of $\frac{6\pi}{7}$, by bringing the bottom edge of the tape UP to coincide with the last crease line sloping up (see Figure 2(c)). We settle for this (because we are content with an odd multiple of π in the numerator) and go to the top of the tape where we see that the angle to the right of the last crease line is $\frac{4\pi}{7}$—and, since we have decided against leaving an even multiple of π in the numerator of any angle next to an edge of the tape, we are forced to bisect this angle twice, each time bringing the top edge of the tape DOWN to coincide with the last crease line, obtaining the arrangement of crease lines shown in Figure 2(d). But we notice that now something miraculous has happened! If we had really started with an angle of $\frac{2\pi}{7}$ and if we now continue introducing crease lines by repeatedly folding DOWN TWICE at the top and UP ONCE at the bottom, we get precisely what we want; namely, pairs of adjacent angles measuring $\frac{\pi}{7}$, at equally spaced intervals along the top edge of the tape. We call this the D^2U^1-folding procedure—or, more simply—and especially when we are concerned merely with the related number theory— the $(2,1)$-folding procedure.

How do we *prove* that this evident convergence takes place? A very direct approach is to assume that the first angle folded down in Figure 2(a) might not have been precisely $\frac{2\pi}{7}$, in which case the bisection forming the next crease would make the two acute angles nearest the top edge of the tape in Figure 2(b) only approximately $\frac{\pi}{7}$; let us call them $\frac{\pi}{7} + \epsilon$ (where ϵ may be either positive or negative). Labeled thus, the angle to the right of the crease, at the bottom of the tape, would measure $\frac{6\pi}{7} - \epsilon$. When this angle is bisected, by folding up, the resulting acute angles nearest the bottom of the tape would both measure $\frac{3\pi}{7} - \frac{\epsilon}{2}$, forcing the angle to the right of this crease line at the top of the tape to measure $\frac{4\pi}{7} + \frac{\epsilon}{2}$. Then when this last angle is bisected twice by folding the tape down, the two acute angles nearest the top edge of the tape will measure $\frac{\pi}{7} + \frac{\epsilon}{2^3}$. This makes it clear that every time we repeat a D^2U^1-folding on the tape the error is reduced by a factor of 2^3.

Notice how our optimistic strategy has paid off. We are in the happy situation that by assuming we have $\frac{\pi}{7}$ to begin with, and proceeding accordingly as described, the angles at the top of the tape get closer and closer to $\frac{\pi}{7}$.

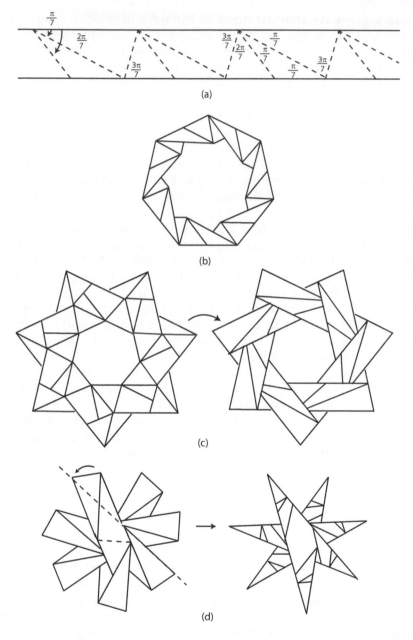

Figure 3. Results of the FAT algorithm: (a) The crease pattern of the folded tape for $\frac{\pi}{7}$. (b) The FAT 7-gon. (c) The FAT $\{\frac{7}{2}\}$-gon. (d) The FAT $\{\frac{7}{3}\}$-gon.

Figure 3(b) shows the FAT 7-gon produced by performing the FAT-algorithm on consecutive vertices along the top of the folded tape shown in Figure 3(a). But, as so often happens in mathematics, we get more than we expected. Figures 3(c) and (d) show the regular $\{\frac{7}{2}\}$- and $\{\frac{7}{3}\}$-gons that are produced from the $(2,1)$-tape by executing the FAT-algorithm on the crease lines that make angles of $\frac{2\pi}{7}$ and $\frac{3\pi}{7}$, respectively, with an edge of the tape (always orient the folded tape so that the angle you are using, for the FAT algorithm, is along the top edge of the tape). In Figures 3(c) and (d) the FAT-algorithm was executed on every other suitable vertex along the edge of the tape so that, in (c), the resulting figure, or its flipped version, could be woven together in a more symmetric way and, in (d), the excess could be folded neatly around the points.

We now want to deal with the case when b is even. We will deal explicitly with the case $b = 14$. Notice that if each angle of $\frac{\pi}{7}$ adjacent to the top edge of the tape in Figure 3(a) is bisected (we call this fold a *secondary fold line*), it will produce two adjacent angles of $\frac{\pi}{14}$ at equally spaced intervals along the top of the tape and we could then fold a FAT 14-gon. By a similar method we could produce 28-gons, 56-gons, etc. Thus, we see that in order to be able to construct an N-gon for any N, all we need to do is to factor N into a power of two times an odd number, say b. Then if we can find a folding procedure that produces a regular b-gon, we can obtain the regular N-gon by introducing secondary fold lines at the appropriate places along the top of the tape. Thus, we need only concern ourselves with finding folding procedures for *odd* numbers b.

We notice, from our example, that the folded tape produced some regular star polygons[3] as well. We will call these regular star polygons $\{\frac{b}{a}\}$-gons, where $a \geq 2$; and we observe that it makes sense to require that $a < \frac{b}{2}$, since to take larger values of a would just result in star polygons where the top edge visits vertices in the opposite order around the bounding b-gon—and each of those star polygons is identical to one already produced. For example, a $\{\frac{7}{5}\}$-gon is the same as a $\{\frac{7}{2}\}$-gon.

Encouraged by the observation that having an angle of $\frac{3\pi}{7}$ on the tape at equally spaced intervals allows us to construct a regu-

---

[3]One may think of the $\{\frac{b}{a}\}$-gon as a polygon where the top edge of the tape visits every ath vertex of the bounding b-gon. For example, we see in Figure 3(c) that for the $\{\frac{7}{2}\}$-gon the top edge visits every second vertex of the bounding 7-gon.

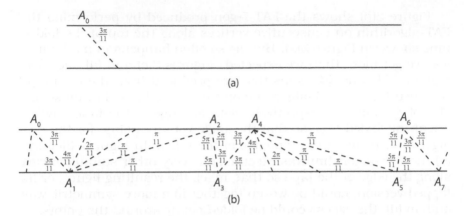

Figure 4. The optimistic strategy for $\frac{3\pi}{11}$: (a) Creasing the initial angle $\frac{3\pi}{11}$. (b) The end result.

lar star $\{\frac{7}{3}\}$-gon, we ambitiously try to construct, by similar means, a regular star $\{\frac{11}{3}\}$-gon. Once again we can obtain the instructions by using our *optimistic strategy*, which means that we assume that we can fold the desired angle of $\frac{3\pi}{11}$, at A_0, in Figure 4(a), and we adhere to the same principles that we used in constructing the regular 7-gon, namely, we adopt the following three rules.

1. Each new crease line goes in the forward (left to right) direction along the strip of paper.

2. Each new crease line always bisects the angle between the last crease line and the edge of the tape from which it emanates.

3. The bisection of angles at any vertex continues until a crease line produces an angle of the form $\frac{a'\pi}{b}$ where a' is an odd number; then the folding stops at that vertex and commences at the intersection point of the last crease line with the other edge of the tape.

Once again the *optimistic strategy* works; and, using the three rules above, we get the tape whose angles converge to those shown in Figure 4(b). We could denote this folding procedure by $D^1U^3D^1U^1D^3U^1$, interpreted in the obvious way on the tape—that

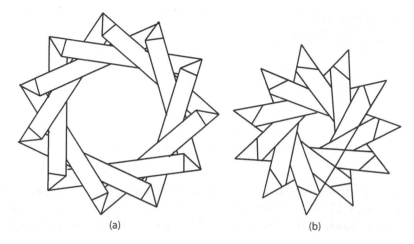

$$\text{(a)} \qquad\qquad\qquad\qquad \text{(b)}$$

Figure 5. (a) The FAT $\{\frac{11}{3}\}$-gon. (b) The FAT $\{\frac{11}{4}\}$-gon.

is, the first exponent "1" refers to the one bisection (producing a crease in a downward direction) at the vertices A_{6n} (for $n = 1, 2, \ldots$) on the top of the tape; similarly the "3" refers to the three bisections at the bottom of the tape (producing creases in an upward direction) made at the vertices A_{6n+1}; etc. However, since the folding is *duplicated* halfway through, we can abbreviate the notation for folding and write simply $(1,3,1)$, with the understanding that we alternately fold from the top and the bottom of the tape as described with the *number* of bisections at each vertex running, in order, through the values 1, 3, 1, ...

The convergence can be shown using an error-correction type of proof similar to that used earlier for the tape that produced the 7-gon. The reader should have no difficulty in supplying the details.

Now consider the tape in Figure 4(b). If the FAT-algorithm is performed on consecutive *parallel* creases of the same length, some $\{\frac{b}{a}\}$-gon will result. If, for example, we use the crease lines $A_{6n}A_{6n+1}$, $0 \le n \le 10$, (all of which make an angle of $\frac{3}{11}\pi$ with the top edge of the tape), we can get the $\{\frac{11}{3}\}$-gon shown in Figure 5(a). Likewise, if we perform the FAT-algorithm on the shortest crease lines emanating downward from A_{6n+4} ($0 \le n \le 10$), we can get the $\{\frac{11}{4}\}$-gon shown in Figure 5(b).

the smallest angle to the right of A_n where	is of the form $\frac{a}{11}\pi$	and the number of bisections at the *next* vertex is x
$n = 0$	$a = 3$	$x = 3$
$n = 1$	$a = 1$	$x = 1$
$n = 2$	$a = 5$	$x = 1$
$n = 3$	$a = 3$	$x = 3$
$n = 4$	$a = 1$	$x = 1$
$n = 5$	$a = 5$	$x = 1.$

Table 1. Conclusions about star 11-gons.

In fact, it is possible to fold, from this tape, *all* of the possible regular star 11-gons;[4] namely, the $\{\frac{11}{1}\}$-, $\{\frac{11}{2}\}$-, $\{\frac{11}{3}\}$-, $\{\frac{11}{4}\}$-, and $\{\frac{11}{5}\}$-gon. Notice that there are fold lines of five different lengths on this tape.

Now, to set the scene for the number theory that comes next, and to enable us to systematically determine the folding procedure for any given a and b, let us look at the patterns in the *arithmetic* of the computations for this last example where $a = 3$ and $b = 11$. Referring to Figure 5(b), we observe that the smallest angle to the right of A_n where is of the form $\frac{a}{11}\pi$ and the number of bisections at the *next* vertex is x for the values of n, a, and x as listed in Table 1.

We could write this as a shorthand *symbol* in the following way:

$$(b =)11 \begin{vmatrix} (a =)3 & 1 & 5 \\ 3 & 1 & 1 \end{vmatrix}. \tag{1}$$

Notice that (1) only has three entries in the top and bottom row (instead of the six you might have expected from the layout above it); this is because we stop when the next a will be the same as the initial number in the top row. We then say that the symbol is *contracted*.

Observe that if we had started with the assumption that our original angle was $\frac{\pi}{11}$, (say, at vertex A_4), we would have gotten identically the same folded tape, but then the symbol would have taken the form

$$(b =)11 \begin{vmatrix} (a =)1 & 5 & 3 \\ 1 & 1 & 3 \end{vmatrix}.$$

---

[4]We will consider the regular 11-gon to be a special regular star polygon denoted, more elaborately, as a $\{\frac{11}{1}\}$-gon.

In fact, it should be clear that we can *start anywhere* (with $a = 1$, 3, or 5), and the resulting symbol analogous to (1) will be obtained by cyclic permutation of the matrix component of the symbol, placing our choice of a in the first position along the top row.

If we followed this procedure in an effort to construct a $\{\frac{33}{9}\}$-gon, we would obtain the symbol

$$(b =)33 \left| \begin{array}{ccc} (a =)9 & 3 & 15 \\ 1 & 1 & 3 \end{array} \right|$$

Notice that this gives precisely the same folding procedure as that for an $\{\frac{11}{3}\}$-gon in the symbol shown in (1). This means that if you fold the $(1, 1, 3)$-tape and perform the FAT-algorithm on the tape on the crease lines making an angle of $\frac{9\pi}{33}$ with the top of the tape, you will obtain an $\{\frac{11}{3}\}$-gon, because, of course, $\frac{33}{9} = \frac{11}{3}$. This is, as we like to say, the most difficult method known to man for reducing fractions. We will avoid this difficulty by requiring that a and b are mutually prime. We may specify this by restricting a so that $\gcd(a, b) = 1$, and we will then call our symbol *reduced*.

In general, if we wish to fold a $\{\frac{b}{a}\}$-gon, with b, a odd and $a < \frac{b}{2}$, then we may construct a symbol as follows. We write

$$b \left| \begin{array}{ccccc} a_1 & a_2 & \cdot & \cdot & a_r \\ k_1 & k_2 & \cdot & \cdot & k_r \end{array} \right| \tag{2}$$

where b, a_i are odd $(a_1 = a)$, $a_i < \frac{b}{2}$, $\gcd(a_1, b) = 1$ (reduced), and

$$b - a_i = 2^{k_i} a_{i+1}, \quad i = 1, 2, \cdots, r, \quad a_{r+1} = a_1 \text{ (contracted)}. \tag{3}$$

Example 1. Suppose we wish to construct all possible star 31-gons. We start by finding instructions for folding a convex 31-gon; thus $b = 31$, $a = 1$, and we construct the symbol

$$(b =)31 \left| \begin{array}{cc} (a =)1 & 15 \\ 1 & 4 \end{array} \right|$$

which tells us that folding $D^1 U^4$ will produce tape that can be used to construct a FAT 31-gon. In fact, we get more. Here are all the FAT polygons that can be constructed from the $D^1 U^4$-tape:

$$\left\{ \frac{31}{1} \right\} -, \left\{ \frac{31}{2} \right\} -, \left\{ \frac{31}{4} \right\} -, \left\{ \frac{31}{8} \right\} -, \text{ and } \left\{ \frac{31}{15} \right\} - \text{gons.}$$

But this folding does not produce a $\{\frac{31}{3}\}$-gon. To obtain the $\{\frac{31}{3}\}$-gon we start with $b = 31$, $a = 3$, obtaining the symbol

$$(b =)31 \left| \begin{array}{cc} (a =)3 & 7 \\ 2 & 3 \end{array} \right.$$

which tells us that by folding D^2U^3—or, more simply, the $(2,3)$-folding procedure—we will produce $(2,3)$-tape from which we can fold the FAT $\{\frac{31}{3}\}$-gon. Again, we get more than we initially sought, the $(2,3)$-tape can be used to construct FAT

$$\left\{\frac{31}{3}\right\} -, \; \left\{\frac{31}{6}\right\} -, \; \left\{\frac{31}{12}\right\} -, \; \left\{\frac{31}{7}\right\} -, \text{ and } \left\{\frac{31}{14}\right\} - \text{gons.}$$

However, we still do not have a procedure for folding a $\{\frac{31}{5}\}$-gon. So, in order to get the $\{\frac{31}{5}\}$-gon, we construct a symbol with $b = 31$, $a = 5$, obtaining

$$(b =)31 \left| \begin{array}{cccc} (a =)5 & 13 & 9 & 11 \\ 1 & 1 & 1 & 2 \end{array} \right.$$

which tells us that by folding $D^1U^1D^1U^2$—or, more simply, by the $(1,1,1,2)$-folding procedure—we produce tape from which we can fold the FAT $\{\frac{31}{5}\}$-gon. Once again, we get bonuses. We see that the $(1,1,1,2)$-tape can be used to construct FAT

$$\left\{\frac{31}{5}\right\} -, \; \left\{\frac{31}{10}\right\} -, \; \left\{\frac{31}{13}\right\} -, \; \left\{\frac{31}{9}\right\} -, \text{ and } \left\{\frac{31}{11}\right\} - \text{gons.}$$

We can combine all the possible symbols for $b = 31$ into one *complete symbol*, calling each part a *coach* (because it looks to us like a coach on a train), with the number of coaches being denoted by c. The complete symbol then takes the form

$$31 \left| \begin{array}{cc|cc|cccc} 1 & 15 & 3 & 7 & 5 & 13 & 9 & 11 \\ 1 & 4 & 2 & 3 & 1 & 1 & 1 & 2 \end{array} \right. \tag{4}$$

Notice that the value of the bottom row sum in each coach of (4) is 5. Notice, too, that the parity in the number of entries is the same in each coach (namely, in this case, even). Is this an accident? The interested reader should try writing out complete symbols for odd numbers of your choice and look for patterns among the numbers involved. The more experience you have with these complete symbols, the better prepared you will be to understand the rest of this article. For example, you may then better appreciate the following theorem concerning every individual coach of the symbol. In the next section we give a theorem concerning the complete symbol. The theorem at hand applies to any coach of a complete symbol.

Theorem 1 (Quasi-order Theorem). *Suppose that a, b are odd, with $a < \frac{b}{2}$, and the symbol*

$$b \begin{vmatrix} a_1 & a_2 & \cdot & \cdot & \cdot & a_r \\ k_1 & k_2 & \cdot & \cdot & \cdot & k_r \end{vmatrix} \tag{5}$$

is obtained using the calculation $b - a_i = 2^{k_i} a_{i+1}$ (with k_i maximal). Let $\sum_{i=1}^{r} k_i = k$, and assume that (5) is not only contracted ($a_{r+1} = a_1$), but also reduced (gcd$(a_i, b) = 1$). Then the quasi-order of $2 \bmod b$ is k. That is, k is the smallest integer such that

$$2^k \equiv \pm 1 \bmod b.$$

In fact, $2^k \equiv (-1)^r \bmod b$. (This means that b exactly divides $2^k - (-1)^r$, which may be written as $b | 2^k - (-1)^r$.)

You may wish to check that the quasi-order theorem is true in the examples you constructed. In the example seen in symbol (1), where $b = 11$, $a = 3$, $r = 3$, and $k = 5$, the quasi-order theorem tells us that $11 | 2^5 - (-1)^3$ or $11 | 32 + 1 = 33$, and that 5 is the *smallest* exponent such that $11 | 2^5 \pm 1$. Again, the quasi-order theorem would tell us for any coach of the symbol (4) where $b = 31$, $k = 5$, and $r = 2$ or 4, that 31 exactly divides $2^5 - 1$, which can hardly be considered a surprise!

The example when $b = 31$ also gives us a glimpse of what is to come. Notice that $\Phi(31) = 30$, $c = 3$, and $k = 5$, so that, in this case $\Phi(b) = 2ck$. Here are a couple of other complete symbols (you should try to construct some of your own) for you to examine to see whether the relationships you have observed are just happy accidents, or whether you believe that they must always happen.

$$43 \begin{vmatrix} 1 & 21 & 11 & 3 & 5 & 19 & 7 & 9 & 17 & 13 & 15 \\ 1 & 1 & 5 & 3 & 1 & 3 & 2 & 1 & 1 & 1 & 2 \end{vmatrix}, \quad \Phi(43) = 42, \quad k = 7, \quad c = 3.$$

$$51 \begin{vmatrix} 1 & 25 & 13 & 19 & 5 & 23 & 7 & 11 \\ 1 & 1 & 1 & 5 & 1 & 2 & 2 & 3 \end{vmatrix}, \ \Phi(51) = 32, \ k = 8, \ c = 2.$$

$$65 \begin{vmatrix} 1 & 3 & 31 & 17 & 7 & 29 & 9 & 11 & 27 & 19 & 23 & 21 \\ 6 & 1 & 1 & 4 & 1 & 2 & 3 & 1 & 1 & 1 & 1 & 2 \end{vmatrix}, \ \Phi(65) = 48, \ k = 6, \ c = 4.$$

The proof of the quasi-order theorem is found in [2–4]. An important tactic used in the proof is to assume that, for a fixed b, S is the set of positive odd numbers $a < \frac{b}{2}$ and then to show that the rule $a \mapsto a'$, (which we call the ψ-algorithm) where

$$b - a = 2^k a',\tag{6}$$

with k *maximal*, is a *permutation* on the finite set S. This proof involves the inverse function, $a' \mapsto a$, (which we call the φ-algorithm) where

$$a = b - 2^k a',\tag{7}$$

with k *minimal* such that $2^k a' > \frac{b}{2}$. It turns out that ψ and φ are mutual *inverses*, and that, not surprisingly, $\gcd(b, a) = \gcd(b, a')$. But the heart of the proof involves the construction, using (7), to work *backward*, of our symbol (5). So, how do we construct the *modified symbol* using (7)?

Recall that the construction of the original symbol involved starting with some number a_1 that we wrote in first position of the top row, then we repeatedly divided $b - a_1$ by 2 until we arrived at an odd number, a_2. (Think of where the original idea came from on the tape before we invented the symbol.) At that point we recorded in the second row under a_1, the number of times, k_1, we divided by 2, and in the top row, to the right of a_1 we recorded a_2. We repeat this process until the next number to be entered in the top row is the same as the original a_1.

In a similar way the construction of the *modified symbol* may be achieved, starting with the same a_1 and repeatedly multiplying a_1 by 2 as many times as it takes, say ℓ_1, until the first time the product is $> \frac{b}{2}$. At that point record in the bottom row slightly to the right of a_1 the number ℓ_1 and record in the top row, slightly to the right of the last entry, the number $b - 2^{\ell_1} a_1$ (this will not be a_2, but rather the last number in the original symbol, which makes sense if this is really the *reverse* algorithm). Continue doing this until the number entered in the top row is again a_1. Thus each coach in a complete symbol has an equivalent modified symbol.

Precisely, in the case of $b = 31$, each coach has a modified symbol as follows:

$$\text{the coach} \begin{vmatrix} 1 & 15 \\ 1 & 4 \end{vmatrix} \text{becomes} \begin{pmatrix} 1 & 15 & 1 \\ & 4 & 1 \end{pmatrix};$$

$$\text{the coach} \begin{vmatrix} 3 & 7 \\ 2 & 3 \end{vmatrix} \text{becomes} \begin{pmatrix} 3 & 7 & 3 \\ & 3 & 2 \end{pmatrix};$$

$$\text{the coach} \begin{vmatrix} 5 & 13 & 9 & 11 \\ 1 & 1 & 1 & 2 \end{vmatrix} \text{becomes} \begin{pmatrix} 5 & 11 & 9 & 13 & 5 \\ & 2 & 1 & 1 & 1 \end{pmatrix}.$$

The Main Theorem

In this section, we enunciate and prove the main theorem of this paper. Let $b > 1$ be an odd number, and let $\Phi(b)$ be the Euler totient function of b. Let us form Σ, the complete symbol of b; and let c be the number of coaches in Σ. Finally, let k be the quasi-order of $2 \bmod b$. Then we will prove Theorem 2

Theorem 2 (Coach Theorem). $\Phi(b) = 2ck$.

You may be interested to look back at the complete symbol for $b = 31$ and verify that this is, indeed, true in that case. Before proving this theorem, we will illustrate it with yet another example.

Example 2. Let $b = 117$. Then, since $b = 9 \cdot 13$, $\Phi(b) = 6 \cdot 12 = 72$. A straightforward calculation shows that Σ, the complete symbol of 117, is given by

$$\Sigma = 117 \begin{vmatrix} 1 & 29 & 11 & 53 & | & 5 & 7 & 55 & 31 & 43 & 37 & | & 17 & 25 & 23 & 47 & 35 & 41 & 19 & 49 \\ 2 & 3 & 1 & 6 & | & 4 & 1 & 1 & 1 & 1 & 4 & | & 2 & 2 & 1 & 1 & 1 & 2 & 1 & 2 \end{vmatrix}. \tag{8}$$

Thus $c = 3$, $k = 12$, and

$$\Phi(117) = \Phi(3^2 \cdot 13) = \Phi(3^2)\Phi(13) = 6 \cdot 12 = 72 = 2 \cdot 3 \cdot 12.$$

We now prove the Coach Theorem.

Proof of Theorem 2: Let $\mathbb{Z}_b^*$ be the multiplicative group of residues mod b prime to b, so that $\text{order}(\mathbb{Z}_b^*) = \Phi(b)$; and let T be the subgroup of $\mathbb{Z}_b^*$ generated by -1 and 2. We first observe that

$$T = \{(-1)^i 2^j \bmod b; \ 0 \leq i \leq 1, \ 0 \leq j \leq k - 1\}, \tag{9}$$

where k is the quasi-order of $2 \bmod b$.

Now a coach is a symbol as defined in [2, page 126]. We also recall here the *modified symbol* introduced in [2, page 131], and recall from the first section how the symbol and the modified symbol are related. In fact, we will obtain a modified symbol from a given element of $\mathbb{Z}_b^*/T$; but it is then an automatic step—which we will describe—to obtain the symbol itself, that is, the coach.

First we show that each element of $\mathbb{Z}_b^*/T$ is represented by a number a that is (i) odd, (ii) prime to b, and (iii) less than $\frac{b}{2}$. We claim that it is obvious from the structure of T that we may represent an element of $\mathbb{Z}_b^*/T$ by a number a' that is (i) odd, (ii) prime to b, and (iii)$'$ less than b. Let a' be such a number. If $a' < \frac{b}{2}$, there is nothing further to do. But if $a' > \frac{b}{2}$, then $b - a' < \frac{b}{2}$ and represents the same element ζ of $\mathbb{Z}_b^*/T$. However $b - a'$ is even, so we may set $b - a' = 2^\ell a$, with a odd and $\ell \geq 1$. Again a represents the same element of $\mathbb{Z}_b^*/T$ as a', and a satisfies conditions (i), (ii), and (iii).

We now apply the reverse φ-algorithm (see [3, page 116] and [4, Equation (4.10)]) to a, and iterate the applications. Thus, writing a_1 for a, we obtain a sequence of odd numbers

$$a_1, a_2, \ldots, a_r, a_{r+1}, \tag{10}$$

all satisfying conditions (i), (ii), and (iii), where

$$a_{r+1} = a_1. \tag{11}$$

Explicitly, the passage from a_1 to a_2 is achieved by repeatedly doubling a_1, until we achieve $2^{\ell_1} a_1 > \frac{b}{2}$, with $\ell_1 \geq 1$ and then set $a_2 = b - 2^{\ell_1} a_1$; and we continue to generate the entire sequence $a_1, a_2, \cdots, a_r$. Plainly $a_1, a_2, \cdots, a_r$ all represent the same element of $\mathbb{Z}_b^*/T$. Now let us insert the even numbers $2a_1, 2^2 a_1, \cdots, 2^{\ell_1} a_1$ between a_1 and a_2, and proceed similarly between a_2 and a_3, ..., a_{r-1} and a_r, a_r and a_1. The result is now precisely the effect of the φ-algorithm on the modified symbol. The φ-algorithm is, of course, inverse to the ψ-algorithm used in the construction of the symbol.

Let us now pause to sum up the steps so far. Starting with a_1 satisfying (i), (ii), and (iii), and representing a given element ζ of $\mathbb{Z}_b^*/T$, we then apply the (reverse) φ-algorithm (see [3, page 116]) to $a = a_1$, and iterate the applications, obtaining (10), namely,

$$a_1, a_2, \ldots, a_r, a_{r+1},$$

all satisfying conditions (i), (ii), and (iii) and with $a_{r+1} = a_1$. Thus, starting with a_1 satisfying conditions (i), (ii), and (iii) and representing a given element ζ of $\mathbb{Z}_b^*/T$, we apply to a_1 the reverse φ-algorithm, that is, we construct the sequence

$$a_1, 2a_1, \ldots, 2^{\ell_1}a_1$$

such that ℓ_1 is minimal for the property $2^{\ell_1}a_1 > \frac{b}{2}$, (so that $\ell_1 \geq 1$), and set

$$a_2 = b - 2^{\ell_1}a_1. \tag{12}$$

We begin again with a_2 and again apply the φ-algorithm to obtain a_3. We continue in this way until we reach $\ldots, a_r, a_{r+1}$, with $a_{r+1} = a_1$. This must occur eventually since φ is a permutation, indeed, the permutation inverse to the ψ-algorithm (see [3, page 116]), which was used to construct the symbol in the first place.

Indeed, it is easy to see how the symbol may be derived from the element of $\mathbb{Z}_b^*/T$ represented by the number a_1 satisfying conditions (i), (ii), and (iii). We write down the sequence arising from the doubling process, together with (12), to pass from a_1 to a_2, and then proceed as described above, obtaining

$$a_1, 2a_1, \ldots, 2^{\ell_1}a_1, a_2, 2a_2, \ldots, 2^{\ell_2}a_2, a_3, \ldots, a_r, 2a_r, \ldots, 2^{\ell_r}a_r, a_1. \tag{13}$$

It is then not difficult to see that we obtain from (13) a coach as follows. We write down the odd terms of (13) as the modified top line of the modified coach, thus

$$a_1, a_2, \ldots, a_r, a_1; \tag{14}$$

and the modified second line of the coach simply lists the number of even numbers between successive odd numbers in the sequence (13). Thus the second line is

$$\ell_1, \ell_2, \ldots, \ell_r. \tag{15}$$

The result is a *modified symbol*, from which the true symbol (or coach) is obtained by omitting the repeated a_1 from the *start* of (14) and then writing each of (14) and (15) backwards. Thus, if we start with the element of $\mathbb{Z}_b^*/T$ represented by the number 1, the sequence (13) is

$$1, 2, 4, 8, 16, 32, 64, 53, 106, 11, 22, 44, 88, 29, 58, 116, 1$$

so that the top line of the modified symbol is

$$1, 53, 11, 29, 1$$

and the second line of the modified symbol is

$$6, 1, 3, 2,$$

which we display together as the modified symbol

$$117 \begin{pmatrix} 1 & 53 & 11 & 29 & 1 \\ & 6 & 1 & 3 & 2 \end{pmatrix}.$$

Thus, finally, the coach is the symbol

$$117 \begin{vmatrix} 1 & 29 & 11 & 53 \\ 2 & 3 & 1 & 6 \end{vmatrix}$$

and, of course, each of the numbers 1, 29, 11, and 53 represents the same element ξ of $\mathbb{Z}_b^*/T$.

It is now plain that, in general, the process thus far described sets up a one-one correspondence between the set of coaches and the elements of $\mathbb{Z}_b^*/T$.

We are now in a position to complete the proof of Theorem 2. We simply have to count the elements in $\mathbb{Z}_b^*/T$. But, of course, $|\mathbb{Z}_b^*| = \Phi(b)$ and $|T| = 2k$, as follows easily from (9). Thus

$$c = |\mathbb{Z}_b^*|\big/|T| = \Phi(b)\big/2k,$$

so

$$\Phi(b) = 2ck. \tag{16}$$

$\square$

Some Corollaries

Of course, the Coach Theorem implies that $\varphi(b)$ is even. However, the standard algorithm for calculating $\varphi(b)$ shows that $\varphi(b)$ is even if b has an odd prime factor, and we have actually assumed that b is odd and $b > 1$. However, it is striking that

$$k \Big| \frac{1}{2}\varphi(b). \tag{17}$$

This follows easily if b is prime, but is not so obvious if b is composite. However, (17) is a trivial consequence of (16).

Let us say that b has a *cyclic coach* if $c = 1$; that is, if we can obtain only one coach from b. We then have Theorem 3.

Theorem 3. b *has a cyclic coach if, and only if,* $\Phi(b) = 2k$, *where* k *is the quasi-order of* 2 mod b.

An example is given by $b = 21$, when $\Phi(b) = 12$, $k = 6$.

Some Open Questions

We leave the reader with some open questions.

1. How can you tell if a complete symbol will have only one coach? How can you predict how many coaches a complete symbol will have?

2. How can you tell what the value of r will be for a given coach?

3. Given b, how are the values of r for different coaches in the complete symbol related?

4. Given b, can you tell which collection of numbers will come together as the top row of a coach?

5. We generalize the symbol from base 2 to an arbitrary base t in Chapter 4 of [3]. Can we generalize the Coach Theorem?

6. For a fixed k, can the sequence of numbers $k_1, k_2, \ldots, k_r$ appear in the bottom row of two distinct coaches? (The answer is no: see [2, page 135]) Are there simple *a priori* necessary and/or sufficient conditions, beyond $\sum k_i = k$, that can tell you whether such a sequence is a bottom row of a coach, short of running the entire computation?

Acknowledgment

The authors would like to thank Nicholas Tran for his help with the technical details of preparing this paper for publication.

Dedication

This article is dedicated to Martin Gardner, who provided the original inspiration for the paper-folding strategy.

Bibliography

[1] Martin Gardner. *Martin Gardner's Mathematical Games: The Entire Collection of his Scientific American Columns*, CD-ROM. Washington, DC: Mathematical Association of America, 2005.

[2] Peter Hilton, Derek Holton, and Jean Pedersen. *Mathematical Reflections: In a Room with Many Mirrors*. New York: Springer-Verlag, 1998.

[3] Peter Hilton, Derek Holton, and Jean Pedersen. *Mathematical Vistas: From a Room with Many Windows*. New York: Springer-Verlag, 2002.

[4] Peter Hilton and Jean Pedersen. *Geometry in Practice and Numbers in Theory*, Monographs in Undergraduate Mathematics 16. Greensboro, NC: Department of Mathematics, Guilford College, 1987.

Seven-Fold Symmetry in Mathematica(l) Graphics and Physical Models

Sándor Kabai

Seven-fold symmetry is explored with the help of graphics generated by the software Mathematica [1]. General procedures are described, with seven-fold symmetry as a particular case. Examples of rhombi and rhombohedra associated with seven-fold symmetry are depicted. A general method of producing rings of $2n$ rhombohedra is introduced, and it is shown how these rings are related to the Yoshimura shape and to a possible spacecraft design. Various methods are introduced for generating spirals. Rolling circles are used to generate various curves. The occurrence of the number seven in two polyhedral structures is described. A variety of shapes, including the torus, heart, submarine, raindrop, barrel, and corn kernel, assume the shape of a polar zonohedron.

Specific Rhombi Associated with Seven-Fold Symmetry

Seven-Fold Tilings

A family of n-fold tilings can be obtained by projecting polar zonohedra orthogonally onto a plane perpendicular to the axis. This

53

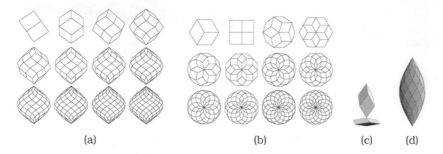

Figure 1. Polar zonohedra for $n = 3, 4, ..., 14$, (a) side views and (b) top views, with three-dimensional depictions of (c) $n = 5$ and (d) $n = 10$.

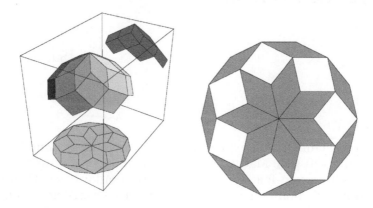

Figure 2. Projections of a polar zonohedron for $n = 7$ (left) and a seven-fold tiling (right).

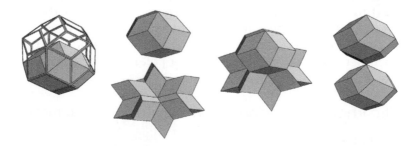

Figure 3. Rhombic icosahedron and ring of ten golden rhombohedra.

axial view is independent of the length scale along the axis (see Figure 1). The number of different tiles required for the tiling depends on n. The $n = 3$, 4, and 5 polar zonohedra are the cube, rhombic dodecahedron, and rhombic icosahedron, respectively. The projections of the $n = 5$ and $n = 10$ polar zonohedra are composed of the two shapes of rhombs that are found in Penrose quasi-periodic tilings.

Mathematica can show three projections of an object simultaneously. Two of these projections of an $n = 7$ polar zonohedron are shown in Figure 2. Only the top part of the polar zonohedron is depicted there. The bottom part is a mirror image of the top part. The tile sizes and vertex positions can be calculated from the axial projection.

Relationship of Rhombohedra and Polar Zonohedra

The rhombic isocahedron is an example of a polar zonohedron. It can be fitted into the "valley" of a ring of ten golden rhombohedra (see Figure 3).

Rings of either rhombic icosahedra or of ten rhombohedra can be constructed. Double rings can also be formed, and such double rings may be stacked to define a cylindrical structure (see Figure 4). Half-rhombohedra may also be used in such an arrangement, and a double helix can be built (see Figure 5).

Let us now try to treat the general case and discover whether analogous structures can be made from rings of $2n$ rhombohedra. To this end we define a tetrahedral component of the rhombohedron. We align two parallel regular n-gons, separated by the

Figure 4. Double ring of rhombic icosahedra (left, middle) and a ring of rhombohedra (right).

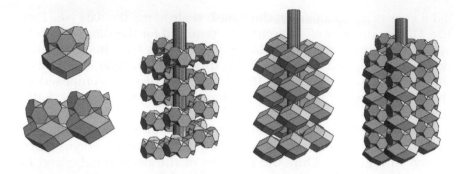

Figure 5. Double helices of rhombic icosahedra and half-rhombohedra forming a sealed cylindrical shape.

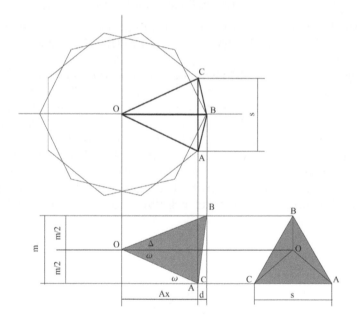

Figure 6. Constructing the tetrahedral component of the rhombohedron.

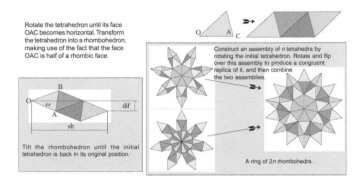

Figure 7. Developing a ring of $2n$ rhombohedra.

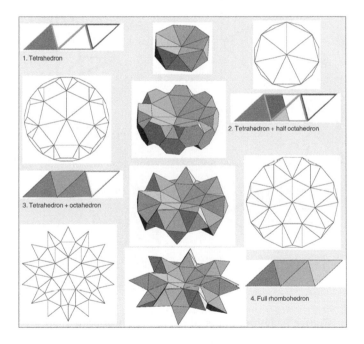

Figure 8 Examples of how the constituent components can be combined.

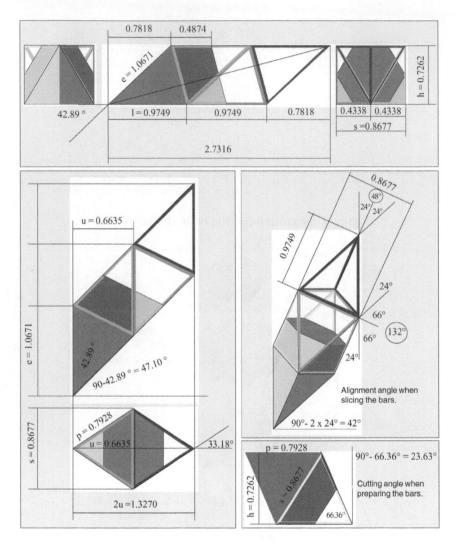

Figure 9. Dimensions of the $n = 7$ rhombohedron.

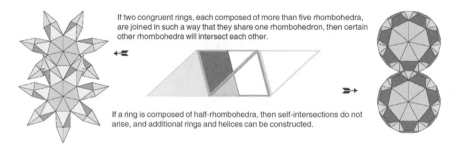

If two congruent rings, each composed of more than five rhombohedra, are joined in such a way that they share one rhombohedron, then certain other rhombohedra will intersect each other.

If a ring is composed of half-rhombohedra, then self-intersections do not arise, and additional rings and helices can be constructed.

Figure 10. Testing the fitting of rhombohedra.

distance m, in such a way that one of them is rotated by π/n relative to the other. We choose the separation distance m so that an edge of one n-gon and a vertex of the other define an equilateral triangle. Figure 6 illustrates the case for $n = 7$. The equilateral triangle and the origin define a tetrahedron (see Figure 6). Figure 7 shows how a ring of $2n$ rhombohedra can be constructed by using this tetrahedron.

Various parts of the rhombohedra can be combined as shown in Figure 8.

The dimensions of the $n = 7$ rhombohedron are shown in Figure 9. Use the following procedure to cut the rhombohedron from a solid material (e.g., EPS or wood). Select a board of thickness t. Calculate the ratio t/h, where h is the height of rhombohedron if the radius of the generating circle is 1, and then multiply all the dimensions in Figure 9 by this ratio. Align the cutting tool at an angle of 23.63°, and cut a bar with rhombic cross-section with rhomb edge length p. Keep the cutting edge at the same angle and slice the bar into rhombohedra while keeping the bar at an angle of 42°.

Let us try to attach two rings in such a way that they share one rhombohedron (see Figure 10).

If n is odd, the diameter of an n-unit ring is larger than that of a helix composed of the same units.

If n is even, the diameter of an n-unit is smaller than that of a helix composed of the same units (see Figure 11). The projection of a helix looks like a ring consisting of $2n$ units. A helix composed of polar zonohedra attached face-to-face can be supplemented with rhombohedral rings. The assembly can be interpreted also as a set of towers consisting of polar zonohedra and rhombohedral rings (see Figure 12).

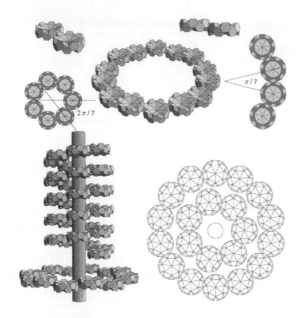

Figure 11. Ring and helix based on $2n$ rhombohedra.

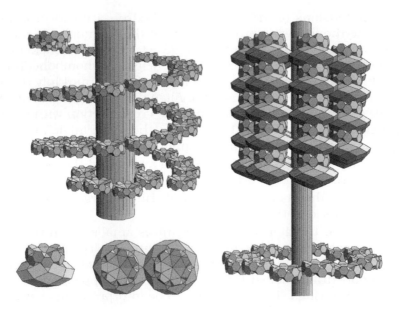

Figure 12. Ring and helix (general case).

Yoshimura Shape

Using the equilateral triangle constructed according to Figure 6, it is possible to form a ring of $2n$ triangles. The Yoshimura shape is obtained by stacking such rings (see Figure 13).

Figures 14 and 15 present the Yoshimura shape for $n = 2, 3, 5, 7$:

$$
\begin{array}{lll}
\text{if } n = 2, & \text{then } m \cong 1.4142 & \text{(tetrahedron);} \\
\text{if } n = 3, & \text{then } m \cong 1.4142 & \text{(octahedron);} \\
\text{if } n = 5, & \text{then } m = 1 & \text{(icosahedron);} \\
\text{if } n = 6, & \text{then } m \cong 0.8556; & \\
\text{if } n = 7, & \text{then } m \cong 0.7449. &
\end{array}
$$

Spacecraft Geometry

It is possible to use the basic arrangement described above to produce new structures:

1. If the point of connection is moved vertically from the origin, then elongated dipyramids are produced (see Figure 16 (left)).

2. If only half of the octahedral component is used, then rings bordered by regular hexagons are obtained. Stacking these rings results in a cylinder bordered by hexagons (see Figure 16 (right).

Recommended "spacecraft" forms include lines, rings, and helices (see Figure 17). The rings can be connected at the hexagonal faces.

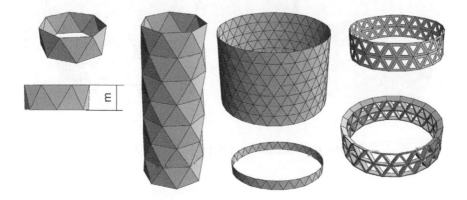

Figure 13. Generation of the Yoshimura shape.

Figure 14. Yoshimura shape for $n = 2, 3, 5, 7$.

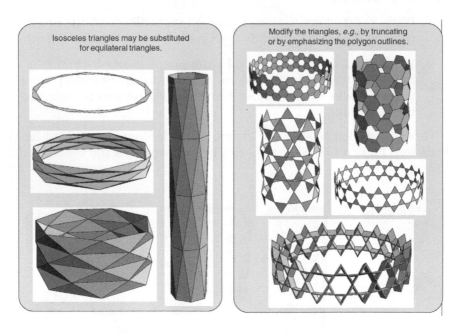

Figure 15. Yoshimura shape for $n = 2, 3, 5, 7$.

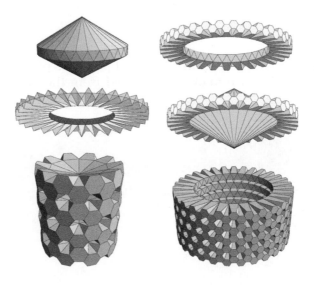

Figure 16. Modifications of the Yoshimura shape.

Figure 17. Formations of rings and helices.

Spirals

Trapezoid Spirals

Select a trapezoid $ABCD$ from a series of polygons plotted as shown in Figure 18. Produce another trapezoid from the original one by

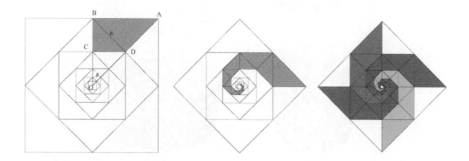

Figure 18. Generating trapezoid spirals.

Figure 19. Trapezoid spirals for $n = 7$.

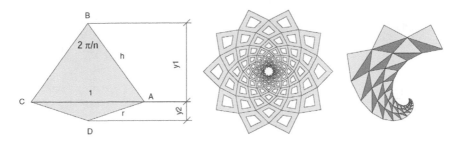

Figure 20. Kite spiral for $n = 10$.

rotating it through the angle β, and enlarging it in the ratio AD/BC. Repeat this procedure to generate the spiral.

Figure 19 shows trapezoid spirals for $n = 7$.

Kite Spirals

Let us construct the kite $(ABCD)$ from two isosceles triangles (see Figure 20) with altitudes $y1$ and $y2$, respectively. The ratio h/r of the sides h and r can be used as the rate of growth for successive kites in a connected sequence.

Figure 21 shows examples of kite spirals for $n = 7$.

Logarithmic Spirals

The vertices of trapezoids and kites in the above spirals are on logarithmic spirals. Additional ways of approximating logarithmic spirals include:

1. *Series of polygons.* Let us plot a connected sequence of n-sided polygons in such a way that the vertices of each *successor* polygon lies on a side of its *antecedent* polygon. This can be accomplished by rotating and increasing the size of the original polygon. The vertices of successive polygons are on logarithmic spirals (see Figure 22(a)). If the angle of rotation exceeds π/n, then the vertices seem to show spirals also in the opposite direction (see Figure 22(b)).

2. *Plotting with parametric equation.* (ParametricPlot3D[{x Sin[x], x Cos[x], 0}, {x, –300, 0, α}]). If the angle of rotation (α) is

Figure 21. Kite spirals for $n = 7$.

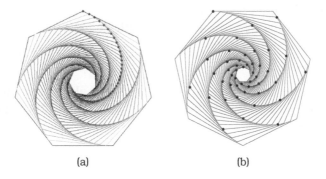

(a)   (b)

Figure 22. Series of polygons, $n = 7$, with angle (a) $\leq \pi/n$ and (b) $> \pi/n$.

$2\pi/\phi$ (where ϕ is the golden ratio), then the number of spirals running left and running right are two adjacent Fibonacci numbers (see Figure 23).

Equilateral triangles can be distorted to create images with n-fold symmetry (see Figure 24).

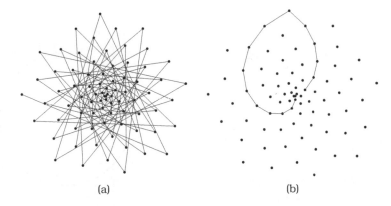

(a)   (b)

Figure 23. Fibonacci spirals.

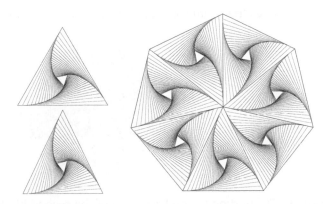

Figure 24. Distorted spirals forming seven-fold symmetry.

Rolling Circles

Suppose we attach points and lines to a circle and roll that circle outside or inside another circle. To obtain seven-fold symmetry, the larger circle must be nine times larger for rolling inside (see Figure 25(a)) and seven times larger for rolling outside (see Figure 25(b)).

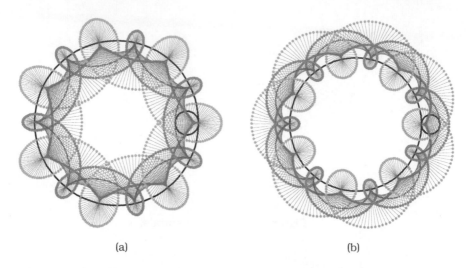

(a)   (b)

Figure 25. Rolling circles: (a) rolling inside and (b) rolling outside.

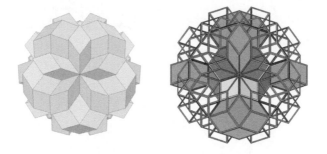

Figure 26. Thirty rhombic dodecahedra of the second kind.

The Number Seven in Polyhedral Structures

Thirty Rhombic Dodecahedra of the Second Kind (RD2)

If a rhombic dodecahedron is placed on each face of a rhombic triacontahedron, seven faces of each RD2 remain exposed, and the resulting assembly has $7 \times 30 = 210$ outside faces (see Figure 26).

Truncated Rhombohedron Module

As can be seen from Figure 9, a rhombohedron can be divided into three major parts; two congruent tetrahedral parts and one octa-

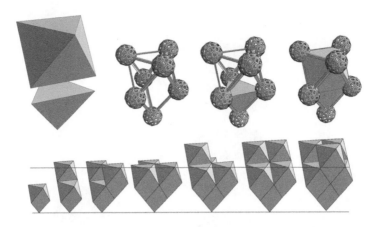

Figure 27. Truncated rhombohedron modules: single module (top left), larger composite modules (bottom), and Zometool module (top left).

hedral part at the center. If one tetrahedral part is combined with the octahedral part (see Figure 27 (top left)), a truncated rhombohedron module is obtained. If we place seven such modules side by side, we get a larger composite module with the same shape (see Figure 27 (bottom)). A truncated rhombohedron of any proportions can be used for this assembly (e.g., the one shown in Figure 8, which can be used to form rings of seven-fold symmetry). If the proportions of the truncated rhombohedron are governed by the golden ratio, rings of five-fold symmetry as well as icosahedral structures can be formed. The module has seven vertices, where truncated rhombic triacontahedra (e.g., Zometool nodes) can be placed and connected with struts along the edges of the module (see Figure 27 (top right)). Note that the struts connect the rectangular faces of the nodes at the top, where the equilateral triangle of the module is located. At all other locations the connections are made at the pentagonal faces of the node.

If the composite module with golden ratio proportions is divided into layers, then planar girders, as well as sphere-like structures with icosahedral symmetry, can be formed (see Figure 28).

If the layer at the middle of the octahedral part of the composite rhombohedron module is selected, then a truncated icosahedron structure can be constructed (see Figure 29).

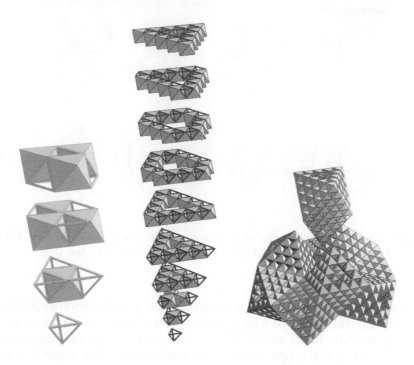

Figure 28. Composite rhombohedron modules cut into layers.

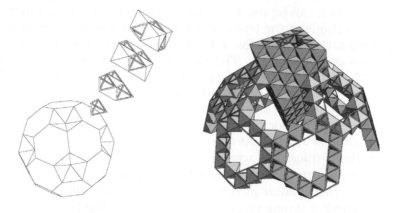

Figure 29. Truncated icosahedral structure made of truncated rhombohedron modules.

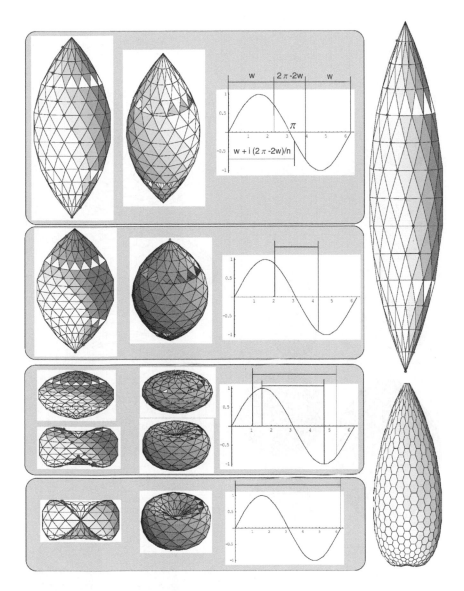

Figure 30. Shaping of polar zonohedra.

Figure 31. Further shaping of polar zonohedra.

Shaping of Polar Zonohedra

First divide each rhomb of the polar zonohedra into two triangles in order to allow a flexible shaping and to ensure that the established shape is bordered by plane units.

Select various ranges of a sine wave to determine the vertical coordinates of the polar zonohedron. Slice the selected range into n parts. Add the coordinate of the ith part to that of the initial point (w) of the range. Select the ranges symmetrically relative to π in order to obtain objects that are symmetrical in the vertical direction (see Figures 30 and 31).

Bibliography

[1] Sándor Kabai. *Mathematical Graphics: Lessons in Computer Graphics Using Mathematica.* Püspökladány, Hungary: Uniconstant, 2002

Seven Knots and Knots in the Seven-Color Map

Louis H. Kauffman

This paper offers two knotty instances of the number seven. The first offering is the remarkable fact, due to John Conway, that any embedding into three-dimensional space of the complete graph on seven nodes will have a knot in it and that this fact is intimately related to knots on the torus and the seven-color map on the torus. We discuss these matters in the first three sections of the paper. The second offering is the equally amazing fact that there are (up to mirror images) exactly seven knots with seven crossings. That is subject of the last section of the paper.

We dedicate this paper to Martin Gardner and to the seventh Gathering for Gardner, held in Atlanta, Georgia in March of 2006.

The Seven-Color Map on the Torus

In Figure 1 you see a knot traced by the bold lines. This knot is topologically equivalent to a simple trefoil knot.

Underneath the knot there is a rectangular bit of hexagonal paving. The paving represents a hexagonal tiling of the surface of

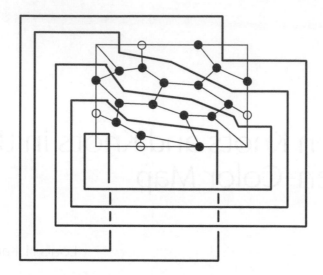

Figure 1. The knot in the seven-color map on the torus.

a torus. The torus is obtained by identifying the top edge of the rectangle with the bottom edge, and the left edge with the right edge. See the later section "The Torus" for more about how the torus is obtained via identifications.

These identifications can be performed in three-dimensional space so that the topology of the knot is preserved and the knot appears on the surface of the torus with the torus embedded in three-dimensional space in the "usual" way. Figure 1 can be interpreted as instructions for drawing a curve (with no self-crossings) on the surface of a torus so that the following two conditions are met:

1. The curve winds around the torus twice in one direction and three times in the other direction, forming a (2,3) torus knot in 3-space (via the standard embedding of the torus in 3-space).

2. The curve goes through each of the seven hexagonal tiles of the tiling of the torus exactly once, meeting the boundaries of the tiles transversely.

The hexagonal tiling of the torus shown in Figure 1 is often called the *seven-color map on the torus*. That is, if we desire to color the regions of the tiling so that two adjacent regions have distinct colors, then seven colors are needed (since each hexagon touches six neighbors).

By other means (using the Euler formula $V - E + F = 0$ for a graph with disc-like regions on the surface of a torus with V nodes, E edges, and F regions), one can show that every map on the torus can be colored with no more than seven colors. The seven-color map shows that seven colors are needed. This yields the *seven-color theorem.*

Theorem 1 (Seven-Color Theorem). *Every map on the torus can be colored with no more than seven colors, and seven is the least number for which this can be stated.*

It is a strange and beautiful fact that the seven-color map on the torus winds upon the surface of the torus forming a kind of vortex when you model it in three dimensions. There is a topological reason for this vortex, and it is related to the knot that we have drawn.

Intrinsic Knotting and the Knot in the Seven-Color Map

A theorem due to Conway [1] says that *the complete graph on seven nodes is intrinsically knotted.* This means that any embedding of K_7 in three-dimensional space must have a knotted cycle. There is a knot in any embedding of K_7 in 3-space.

A complete graph on a set of nodes is a graph in which there is exactly one edge between each pair of possible nodes. See Figure 2 for a depiction of the complete graphs K_1 to K_7.

In Figure 2 you see a specific embedding of K_7 in 3-space. *We leave it to you to find the knot!* The knot will be in the form of a walk along the edges of K_7 (not all of them!) that goes through each node once. (If you could find such a walk that uses less than all of the nodes, that would be fine so long as it is knotted in 3-space. In this example there is no such walk.) In K_6 you will be able to find two curves that are linked with one another. This is a related theorem.

Theorem 2. *The graph K_6 is intrinsically linked.*

It is much easier to find the link in K_6 as shown in Figure 2, and in fact we have illustrated it in that figure with bold lines.

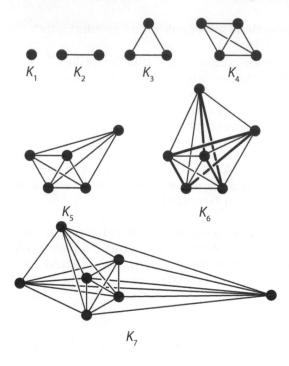

Figure 2. Complete graphs.

The key point about these theorems concerning the intrinsic linkedness of K_6 and the intrinsic knottedness of K_7 is that the linking or knotting will occur *no matter how the graph is embedded in 3-space.* Thus you can make a drawing yourself of K_6 or K_7, and it is guaranteed that no matter how you draw it, how you set the self-crossings, how you make the connections in 3-space, there will be a link in K_6 and there will be a knot in K_7!

Now seven nodes, each of which is connected to every other node, are like seven hexagons, each of which shares a boundary with every other hexagon. The fact that there is a paving of the torus with seven hexagons tells us that there is a drawing of the graph K_7 in the surface of the torus with one node for each hexagon and one graphical edge for each boundary edge of a hexagon.

So by the Conway theorem there has to be a knot inside the K_7 that is embedded in the torus (in 3-space) that is determined by the seven-color map on the torus. A knot that is induced from a curve on a torus must wind around the torus. The simplest such knot winds around the torus three times in the meridian direction on the torus and twice in the longitude direction. This means that *it is not an accident that the hexagonal paving on the torus winds around it.* The winding is needed for the Conway theorem to be true. This is the topology of the seven-color map.

The Torus

We take a rectangle and identify the opposite sides.

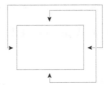

This gives a torus.

Suppose you cut a hole in the surface of the torus.

Now enlarge the hole.

Flatten the remaining material and you find that you have two annuli attached to one another. This can be described by saying that you have a (black) rectangle with the top and bottom attached by a strip and the left and right edges attached by another strip, as shown in Figure 3. That picture is therefore topologically a torus with a hole removed.

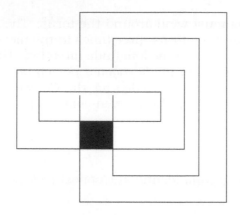

Figure 3. A punctured torus.

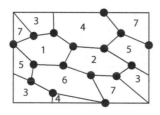

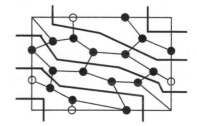

Figure 4. The seven-color map on
the torus.

Figure 5. The seven-color map on
the torus.

Figure 4 shows the seven-color map on the torus, depicted by
thinking of the torus as a rectangle with opposite edges identified.
The seven regions are labeled 1, 2, ..., 7. Note that after making the
identifications of the opposite sides there is one hexagonal region
for each integer from 1 to 7.

Figure 5 adds the drawing of the curve on the torus that becomes our trefoil knot.

Finally, Figure 6 shows the trefoil knot as it appears on the
torus itself.

Or maybe you would prefer the depiction in Figure 7 (which
along with being shaded and rendered is the mirror image of the
trefoil in Figure 6).

In Figure 8, there is another, more vivid depiction of the seven-
color map on the torus, due to Stan Tenen [3].

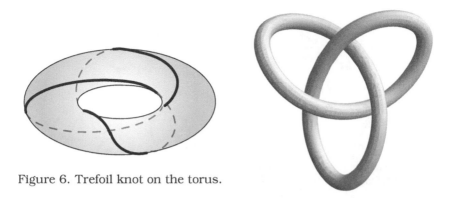

Figure 6. Trefoil knot on the torus.

Figure 7. A trefoil knot.

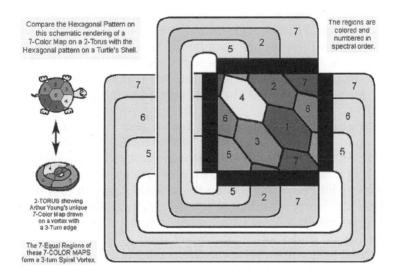

Figure 8. The seven-color map on the torus. (See Color Plate VIII.)

Seven Knots of Seven Crossings

Our second offering of seven is the fact that there are seven knots (up to mirror images) with seven crossings. These are illustrated in Figure 9.

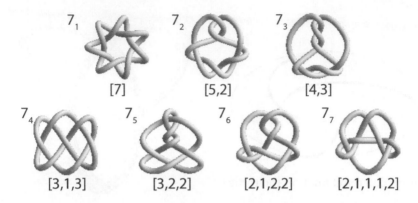

Figure 9. The seven knots with seven crossings.

It is a consequence of a theorem of Schubert (see [2]) that the so-called rational knots and links of n crossings are (up to mirror images) in one-to-one correspondence with ordered partitions of the number n, taken up to completely reversing the order and such that the beginning number and the end number of the partition are not equal to one. Up through seven crossings, all knots are rational.

Given a partition as described above, one associates a continued fraction with it. For example we, associate $3 + 1/(1 + 1/3)$ to the partition $[3, 1, 3]$ of 7. Some partitions will give links rather than knots. It is a fact that the associated continued fraction will give a knot exactly when the corresponding fraction has an odd numerator. Thus $[2, 2]$ corresponds to $2 + 1/2 = 5/2$ and gives a knot, while $[4]$ corresponds to $4 = 4/1$ and gives a link. Figure 9 illustrates only the knots, and there are seven of them.

We leave it as an exercise for the reader to list all the partition classes for $n = 7$ and to see that we have not missed any of the rational knots with seven crossings.

What does this curious relationship with continued fractions and partitions mean? Figure 10 suggests the answer.

In Figure 10, we show how to associate a knot or link to an ordered partition or its associated continued fraction. First one makes a tangle (a weave with four free ends) and then closes this weave to form the corresponding knot or link. The weave has successive twists that correspond to the terms in the partition so that the total number of crossings in the knot or link is equal to the

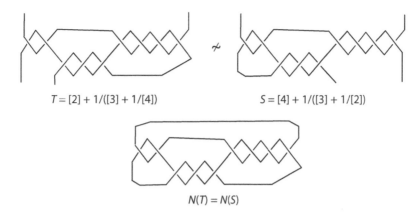

$T = [2] + 1/([3] + 1/[4])$ $S = [4] + 1/([3] + 1/[2])$

$N(T) = N(S)$

Figure 10. Associating a knot or link to a continued fraction.

sum of the terms of the partition. Then one closes the tangle to from the knot or link. In Figure 10 we have indicated two tangles T and S and their closures $N(T)$ and $N(S)$. The two tangles correspond respectively to the partitions $[2, 3, 4]$ and $[4, 3, 2]$. They close to the same knot. We have shown this weaving pattern when the partition has an even number of terms. We leave it to the reader to puzzle out the corresponding pattern for an odd number of terms, and we suggest that the reader make his or her own diagrams of each of the seven knots of seven crossings from their corresponding partitions.

Bibliography

[1] John H. Conway and Cameron McA. Gordon. "Knots and Links in Spatial Graphs." *Journal of Graph Theory* 7 (1983), 445–453.

[2] Louis H. Kauffman and Sofia Lambropoulou. "On the Classification of Rational Knots." *L'Enseignement Mathematiques* 49 (2003), 357–410.

[3] Stan Tenen. Meru Foundation homepage. http://www.meru.org/, 2008.

Seven from the Sea

Michael S. Longuet-Higgins

Traditionally the world has seven seas. But the number seven is involved with the oceans in a much more intimate way. Some marine diatoms—tiny animals that float freely throughout the oceans in large numbers—have the form of a cylindrical drum (Figure 1). The two circular faces, which grow outwards from the central axis, each consist of an array of circular pores. Each pore is surrounded generally by six neighboring pores, so that the pattern resembles a hexagonal honeycomb. The single exception is the central pore, which has seven neighbors. This has the effect of causing the radius of the other pores to diminish with increasing distance from the central axis, with certain physiological consequences.

Can we model the distorted array so as to find the rate of decrease of the pore radius?

Consider a purely hexagonal array as in Figure 2. Let us take polar coordinates (r, θ) with the origin at the center of one of the pores and write

$$z = re^{i\theta} \tag{1}$$

for the complex variable z. Now make the transformation to a new variable, for example

Figure 1. Micrograph of a marine diatom (*Thalassiosira oestrupii*) showing the central pore, or *areola*, with seven neighbors. The horizontal width is 23μm.

$$\zeta = z^{6/7} = \rho e^{i\phi}. \tag{2}$$

This transformation, being conformal, transforms each circular arc into a circular arc, and gives us the pattern in Figure 3. Moreover, from (1) and (2) we have

$$\rho = r^{6/7}, \qquad \phi = \frac{6}{7}\theta. \tag{3}$$

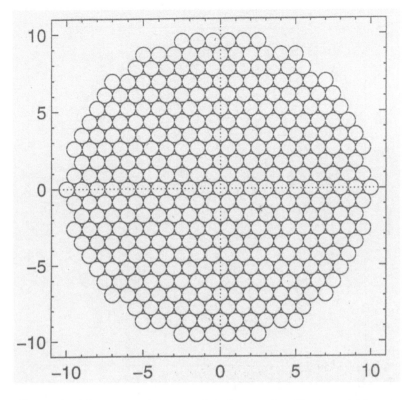

Figure 2. A honeycomb array of circles, each with six neighbors.

As the unit of length, take the diameter of each circle in Figure 2. Then, in Figure 3 the diameter D_0 of the central circle is

$$D_0 = 2 \times \left(\frac{1}{2}\right)^{6/7} = 2^{1/7}, \tag{4}$$

and the diameter D_n of the nth circle from the center is given by

$$D_n = \left(n + \frac{1}{2}\right)^{6/7} - \left(n - \frac{1}{2}\right)^{1/2}, \qquad n > 0. \tag{5}$$

For large n we can approximate D_n by

$$D_n \sim \frac{d}{dn} n^{6/7} = \frac{6}{7} n^{-1/7} \tag{6}$$

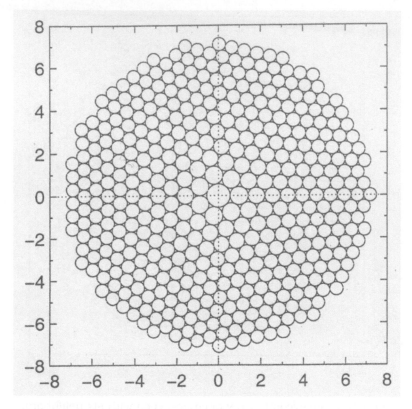

Figure 3. The honeycomb of Figure 2 after transformation by $\zeta = z^{6/7}$.

so that from equations (4) and (6) we find

$$D_n/D_0 \sim \frac{6}{7}(2n)^{-1/7}. \tag{7}$$

A comparison of D_n/D_0 as given by the asymptotic formula (7) with the exact values given by equations (4) and (5) is shown in Figure 4. From this comparison it is clear that equation (7) is a good approximation. For large n we have also

$$r \sim n, \quad \rho \sim n^{6/7} \tag{8}$$

so that

$$D_n/D_0 \sim \frac{6}{7}(2\rho)^{-1/6}. \tag{9}$$

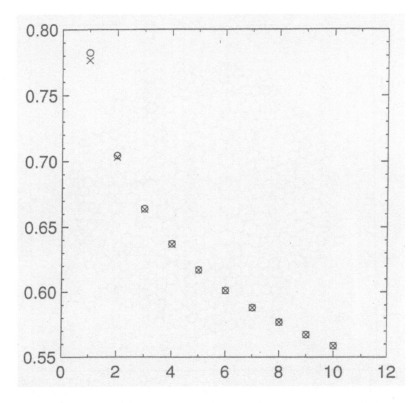

Figure 4. The diameter D_n of the nth pore from the center of the array, as a function of n. Circular plots give exact values; crosses correspond to equation (7).

In other words, the diameter of the pores in Figure 1 must diminish like the one-sixth power of the distance from the central axis.

What if the central pore had more neighbors, say m neighbors? Then by making the transformation

$$\zeta = z^{6/m} \tag{10}$$

in Figure 2 and following exactly the same argument, we arrive at the result that

$$D_n/D_0 \sim \frac{6}{m}(2\rho)^{(6-m)/6}. \tag{11}$$

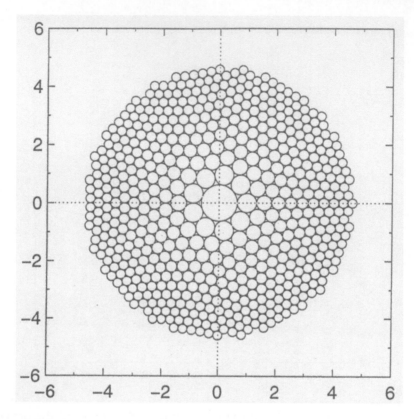

Figure 5. A honeycomb array in which the central pore has nine neighbors.

When $m = 7$ we retrieve equation (8). But if $m = 9$, say, we find

$$D_n/D_0 \sim \frac{2}{3}(2\rho)^{-1/2} \qquad (12)$$

so that the pore size decreases more rapidly than with $m = 7$ (see Figure 5).

What if m is less than 7? With $m = 6$, of course, the rate of decay is zero; the pore size is constant. But if $m < 6$, the pore size *increases* with radial distance. For example with $m = 5$, equation (11) gives

$$D_0/D_0 \sim \frac{6}{5}(2\rho)^{1/6} \qquad (13)$$

as can be seen in Figure 6.

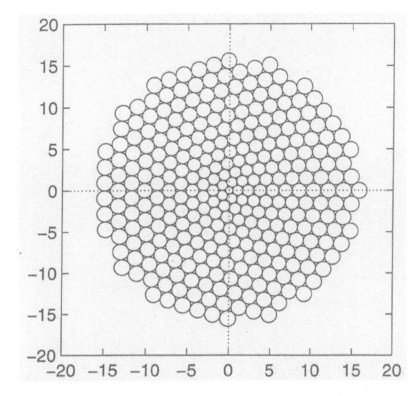

Figure 6. A honeycomb array in which the central pore has five neighbors.

The marine diatoms in which such arrays of circular pores are found are known as *thalassiosira* (from the Greek *thalassa* meaning *sea*), and the particular species in which the seven-fold central pore is found is *thalassiostra oestrupii* after its discoverer, named Westrup. In other species of thalassiosira, we find other types of geometrical constraints arising from the boundaries of the array. The resulting distortions can be treated by other types of conformal transformation. Further details may be found in [1].

Bibliography

[1] M. S. Longuet-Higgins. "Geometrical Constraints on the Development of a Diatom." *J. Marine Biol.* 210 (2001), 101–105.

Seven Staggering Sequences

N. J. A. Sloane

When the *Handbook of Integer Sequences* came out in 1973, Philip Morrison gave it an enthusiastic review in *Scientific American* and Martin Gardner was kind enough to say in his "Mathematical Games" column for July 1974 that "every recreational mathematician should buy a copy forthwith." That book contained 2,372 sequences. Today the *On-Line Encyclopedia of Integer Sequences* (or OEIS) [25] contains over 140,000 sequences. The following are seven that I find especially interesting. Many of them quite literally stagger. The sequences will be labeled with their numbers (such as A064413) in the OEIS. Much more information about them can be found there and in the references cited.

The EKG Sequence

(This is A064413 and is due to Jonathan Ayres.) The first three sequences are defined by unusual recurrence rules. The first begins with $a(1) = 1$, $a(2) = 2$, and the rule for extending it is that the next term, $a(n+1)$, is taken to be the smallest positive number not already in the sequence that has a nontrivial common factor with the previous term $a(n)$. Since $a(2) = 2$, $a(3)$ must be even,

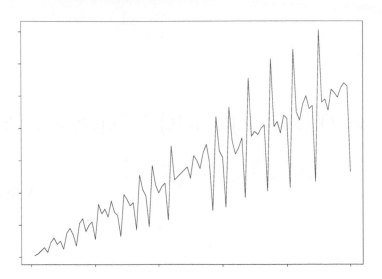

Figure 1. The first 100 terms of the EKG sequence, with successive points joined by lines.

and is therefore 4; $a(4)$ must have a factor in common with 4, that is, must also be even, and so $a(4) = 6$. The smallest number not already in the sequence that has a common factor with 6 is 3, so $a(5) = 3$, and so on. The first 18 terms are

$$1, 2, 4, 6, 3, 9, 12, 8, 10, 5, 15, 18, 14, 7, 21, 24, 16, 20, \dots.$$

It is clear that if a prime p appears in the sequence, $2p$ will be the term either immediately before or after it. Jeffrey Lagarias, Eric Rains, and I have studied this sequence [19]. One of the things that we observed was that in fact every odd prime p was always *preceded* by $2p$, and always *followed* by $3p$. This is certainly true for the first 10,000,000 terms, but we were unable to prove it in general. (This has recently been established by Piotr Hofman and Marcin Pilipczuk [16].)

We called this the EKG sequence, since it looks like an electro-cardiogram when plotted (Figures 1 and 2).

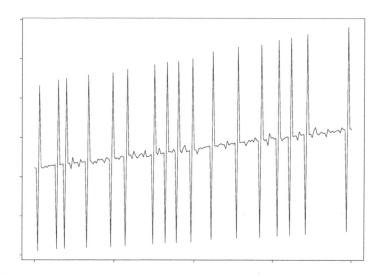

Figure 2. Terms 800 to 1,000 of the EKG sequence.

There is an elegant three-step proof that every positive number must eventually appear in the sequence. (i) If infinitely many multiples of some prime p appear in the sequence, then every multiple of p must appear. (For if not, let kp be the smallest missing multiple of p. Every number below kp either appears or it doesn't, but once we get to a multiple of p beyond all those terms, the next term must be kp, which is a contradiction.) (ii) If every multiple of a prime p appears, then every number appears. (The proof is similar.) (iii) Every number appears. (For if there are only finitely many different primes among the prime factors of all the terms, then some prime must appear in infinitely many terms, and the result follows from (i) and (ii). On the other hand, if infinitely many different primes p appear, then there are infinitely many numbers $2p$, as noted above, so 2 appears infinitely often, and again the result follows from (i) and (ii).)

Although the initial terms of the sequence stagger around, when we look at the big picture we find that the points lie very close to three almost-straight lines (Figure 3). This is somewhat similar to the behavior of the prime numbers, which are initially erratic, but

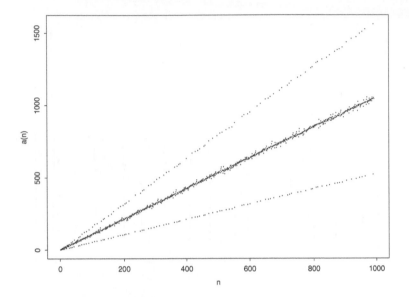

Figure 3. The first 1,000 terms of the EKG sequence, successive points not joined. They lie roughly on three almost-straight lines.

lie close to a smooth curve (since the nth prime is roughly $n \log n$) when we look at the big picture—see Don Zagier's lecture on "The First 50 Million Prime Numbers" [29].

In fact, we have a precise conjecture about the three lines on which the points lie. We believe—but were unable to prove—that almost all $a(n)$ satisfy the asymptotic formula $a(n) \sim n(1+1/(3 \log n))$ (the central line in Figure 3), and that the exceptional values $a(n) = p$ and $a(n) = 3p$, for p a prime, produce the points on the lower and upper lines. We *were* able to show that the sequence has essentially linear growth (there are constants c_1 and c_2 such that $c_1 n < a(n) < c_2 n$ for all n), but the proof of even this relatively weak result was quite difficult. It would be nice to know more about this sequence!

Gijswijt's Sequence

(This is A090822 and was invented by Dion Gijswijt when he was a graduate student at the University of Amsterdam, and analyzed by him, Fokko van de Bult, John Linderman, Allan Wilks, and myself [3].) We begin with $b(1) = 1$. The rule for computing the next term, $b(n+1)$, is again rather unusual. We write the sequence of numbers we have seen so far,

$$b(1), b(2), \ldots, b(n),$$

in the form of an initial string X, say (which can be the empty string $\varnothing$), followed by as many repetitions as possible of some nonempty string Y. That is, we write

$$b(1), b(2), \ldots, b(n) \;=\; XY^k, \text{ where } k \text{ is as large as possible.} \qquad (1)$$

Then $b(n+1)$ is k.

Some examples will make this clear. The sequence begins:

$$1, 1, 2, 1, 1, 2, 2, 2, 3, 1, 1, 2, 1, 1, 2, 2, 2, 3, 2, 1, 1, 2, \ldots.$$

After the first six terms we have

$$b(1), b(2), \ldots, b(6) = 1, 1, 2, 1, 1, 2,$$

so we can take X to be empty, Y to be $1, 1, 2$ and $k = 2$, so $b(1)$, $b(2), \ldots,\ b(6) = Y^2$. This is the largest k we can achieve here, so $b(7) = 2$. Now we have

$$b(1), b(2), \ldots, b(7) = 1, 1, 2, 1, 1, 2, 2,$$

and we can take $X = 1, 1, 2, 1, 1$, $Y = 2$, $k = 2$, getting $b(8) = 2$. Next,

$$b(1), b(2), \ldots, b(8) = 1, 1, 2, 1, 1, 2, 2, 2,$$

and we can take $X = 1, 1, 2, 1, 1$, $Y = 2$, $k = 3$, getting $b(9) = 3$, the first time a 3 appears. And so on.

The first time a 4 appears is at $b(220)$. We computed several million terms without finding a 5, and for a while we wondered if perhaps no term greater than 4 was ever going to appear. However, we were able to show that a 5 does eventually appear, although the universe would grow cold before a direct search would find it. The first 5 appears at about term

$$10^{10^{23}}.$$

The sequence is in fact unbounded, and the first time that a number m $(= 5, 6, 7, \ldots)$ appears is at about term number

$$
2^{2^{3^{4^{\cdot^{\cdot^{\cdot^{m-1}}}}}}},
$$

a tower of height $m - 1$.

There are of course several well-known sequences that have an even slower growth rate than this one (the inverse Ackermann function [1], the Davenport-Schinzel sequences [24], or the inverse to Harvey Friedman's sequence [9], for example). Nevertheless, I think the combination of slow growth and an unusual definition make Gijswijt's sequence remarkable. It also has an interesting recursive structure, which is the key to its analysis. There is only room here to give a hint of this.

The starting point is the observation that the sequence—let's call it $A^{(1)}$—can be built up recursively from "blocks" that are always doubled and are followed by "glue" strings. The first block is $B_1 = 1$, the first glue string is $S_1 = 2$, and the sequence begins with

$$
B_1 B_1 S_1 \; = \; 1, 1, 2,
$$

which is the second block, B_2. The second glue string is $S_2 = 2, 2, 3$, and the sequence also begins with

$$
B_2 B_2 S_2 \; = \; 1, 1, 2, 1, 1, 2, 2, 2, 3,
$$

which is the third block, B_3. This continues: for all m, the sequence begins with $B_{m+1} = B_m B_m S_m$, where S_m contains no 1's and is terminated by the first 1 that follows $B_m B_m$. Now something remarkable happens. If we concatenate all the glue strings $S_1, S_2, S_3, \ldots$, we get a new sequence, $A^{(2)}$ say:

$$
2, 2, 2, 3, 2, 2, 2, 3, 2, 2, 2, 3, 3, 2, 2, 2, 3, 2, 2, 2,
$$
$$
3, 2, 2, 2, 3, 3, 2, 2, 2, 3, 2, 2, 2, 3, 2, 2, 2, 3, 3, 3, 3, \ldots
$$

(A091787), which turns out to be generated by the same rule, (1), as the original sequence, except that the next term is now the maximum of k and 2. If $k = 1$ is the best we can achieve, we promote it to 2. We call $A^{(2)}$ the *second-order* sequence. This has a similar recursive structure as the original sequence, only now it is built up from blocks that are repeated three times and followed by second-order glue strings that contain no 1's or 2's. If we concatenate the second-order glue strings, we get the *third-order* sequence $A^{(3)}$, which is built up from blocks that are repeated four times and followed by third-order glue strings that contain no 1's, 2's, or 3's. And so on.

Now we observe that arbitrarily long initial segments of the second-order sequence $A^{(2)}$ appear as subsequences of the original sequence $A^{(1)}$, and arbitrarily long initial segments of the third-order sequence $A^{(3)}$ appear as subsequences of the second-order sequence $A^{(2)}$, ... But the mth-order sequence $A^{(m)}$ begins with m. So the original sequence contains every positive number!

Of course all this requires proof, and the reader is referred to [3] for further information about this fascinating sequence.

We observed experimentally that in variations of Gijswijt's sequence with initial conditions consisting of any finite string of 2's and 3's, a 1 always eventually appeared in the sequence, but were unable to prove that this would always be the case. We called this the "Finiteness Conjecture": start with any finite initial string of numbers, and extend it by the "next term is k" rule (1). Then eventually one must see a 1. If we had a direct proof of this, it would simplify the analysis of the original sequence. Can some reader find a proof?

Numerical Analogs of Aronson's Sequence

Aronson's sentence is a classic self-referential assertion: "t is the first, fourth, eleventh, sixteenth, ... letter in this sentence" [2, 17] and produces the sequence $1, 4, 11, 16, 24, 29, 33, 35, 39, 45, \ldots$ (A005224). It suffers from the drawback that later terms are ill-defined, because of the ambiguity in the English names for numbers—some people say "one hundred and one," others "one hundred one," etc.

Another well-known self-referential sequence is Golomb's sequence, which is defined by the property that the nth term is the number of times n appears in the sequence

$$1, 2, 2, 3, 3, 4, 4, 4, 5, 5, 5, 6, 6, 6, 6, 7, 7, 7, 7, 8, \ldots$$

(A001462). There is a simple formula for the nth term: it is the nearest integer to (and approaches)

$$\phi^{2-\phi} n^{\phi-1},$$

where $\phi = (1 + \sqrt{5})/2$ is the golden ratio ([10], [12, Section E25]).

In [4], Benoit Cloitre, Matthew Vandermast, and I studied some new kinds of self-referential sequences, one of which is as follows: $c(n + 1)$ is the smallest positive number $> c(n)$ consistent with the condition "n is a member of the sequence if and only if $c(n)$ is odd."

What is $c(1)$? Well, 1 is the smallest positive number consistent with the conditions, so $c(1)$ must be 1. What about $c(2)$? It must be at least 2, and it can't be 2, for then 2 would be in the sequence, but $c(2)$ is even. Nor can it be 3, for then 2 would be missing (the sequence increases) whereas $c(2)$ would be odd. But $c(2)$ *could* be 4, and therefore *must* be 4. So $c(3)$ must be even and > 4, and $c(3) = 6$ works. Now 4 is in the sequence, so $c(4)$ must be odd, and $c(4) = 7$ works. Continuing in this way we find that the first few terms are as follows (this is A79000):

n :	1	2	3	4	5	6	7	8	9	10	11	12	$\cdots$
$c(n)$:	1	4	6	7	8	9	11	13	15	16	17	18	$\cdots$.

Once we are past $c(2)$ there are no further complications, $c(n-1)$ is greater than n, and we *can*, and therefore *must*, take

$$c(n) = c(n-1) + \epsilon,$$

where ϵ is 1 or 2 and is given by

	$c(n-1)$ even	$c(n-1)$ odd
n in sequence	1	2
n not in sequence	2	1.

The gap between successive terms for $n \geq 3$ is either 1 or 2.

The analogy with Aronson's sequence is clear. Just as Aronson's sentence indicates exactly which of its terms are t's, $\{c(n)\}$ indicates exactly which of its terms are odd.

It is easy to show that all odd numbers ≥ 7 occur in the sequence. For suppose some number $2t + 1$ was missing. Therefore $c(i) = 2t$, $c(i+1) = 2t + 2$ for some $i \geq 3$. From the definition, this means i and $i+1$ are missing, implying a gap of at least 3, a contradiction to what we just observed.

Table 1 shows the first 72 terms, with the even numbers underlined.

n:	1	2	3	4	5	6	7	8	9	10
$c(n)$:	1	4	6	7	8	9	11	13	15	16

n:	11	12	13	14	15	16	17	18	19	20
$c(n)$:	17	18	19	20	21	23	25	27	29	31

n:	21	22	23	24	25	26	27	28	29	30
$c(n)$:	33	34	35	36	37	38	39	40	41	42

n:	31	32	33	34	35	36	37	38	39	40
$c(n)$:	43	44	45	47	49	51	53	55	57	59

n:	41	42	43	44	45	46	47	48	49	50
$c(n)$:	61	63	65	67	69	70	71	72	73	74

n:	51	52	53	54	55	56	57	58	59	60
$c(n)$:	75	76	77	78	79	80	81	82	83	84

n:	61	62	63	64	65	66	67	68	69	70
$c(n)$:	85	86	87	88	89	90	91	92	93	95

n:	71	72	$\cdots$
$c(n)$:	97	99	$\cdots$

Table 1. The first 72 terms of the sequence "n is in the sequence if and only if $c(n)$ is odd."

Examining the table, we see that there are three consecutive numbers, 6, 7, 8, which are necessarily followed by three consecutive odd numbers, $c(6) = 9$, $c(7) = 11$, $c(8) = 13$. Thus 9 is present, 10 is missing, 11 is present, 12 is missing, and 13 is present. Therefore the sequence continues with $c(9) = 15$ (odd), $c(10) = 16$ (even), ..., $c(13) = 19$ (odd), $c(14) = 20$ (even). This behavior is repeated forever. A run of consecutive numbers is immediately followed by a run of the same length of consecutive odd numbers. And a run of consecutive odd numbers is immediately followed by a run of twice that length of consecutive numbers (alternating even and odd). Once we have noticed this, it is straightforward to find an explicit formula that describes this sequence:

$$c(1) = 1, \quad c(2) = 4,$$

and subsequent terms are given by

$$c(9 \cdot 2^k - 3 + j) = 12 \cdot 2^k - 3 + \frac{3}{2}j + \frac{1}{2}|j|$$

for $k \geq 0$, $-3 \cdot 2^k \leq j < 3 \cdot 2^k$ (see [4] for the proof).

The structure is further revealed by examining the sequence of first differences, $c'(n) = c(n+1) - c(n)$, which is

$$3, 2, 1, 1, 1, 2, 2, 2, 1^6, 2^6, 1^{12}, 2^{12}, 1^{24}, 2^{24}, \dots,$$

where we have written 1^6 to indicate a run of six 1's, etc. The oscillations double in length at each step.

Approximate Squaring

The symbol $\lceil x \rceil$ denotes the *ceiling function*, the smallest integer greater than or equal to x. Start with any fraction greater than 1, say $\frac{8}{7}$, and repeatedly apply the "approximate squaring" map:

$$\text{replace } x \text{ by } x\lceil x \rceil. \tag{2}$$

Since $\frac{8}{7} = 1.142\dots$, $\lceil \frac{8}{7} \rceil = 2$, so after the first step we reach $\frac{8}{7} \times 2 = \frac{16}{7}$. A second approximate squaring step takes us to $\frac{16}{7} \times 3 = \frac{48}{7}$, and a third step takes us to $\frac{48}{7} \times 7 = 48$, which is an integer, and we stop. It took three steps to reach an integer. The question is: *do we always reach an integer?* Jeffrey Lagarias and I studied this problem in [20]. We showed that almost all initial fractions greater than 1 eventually reach an integer, and that if the denominator is 2 then they all do, but we were unable to give an affirmative answer in general. In fact, we show that the problem has some similarities to the notorious Collatz (or "$3x + 1$") problem [18], and so may be difficult to solve in general. (A similar problem has been posed by Jim Tanton [27].)

The numbers involved grow very rapidly: if we start with $\frac{6}{5}$, for example, successive approximate squarings produce the sequence

$$\frac{6}{5}, \frac{12}{5}, \frac{36}{5}, \frac{288}{5}, \frac{16704}{5}, \frac{55808064}{5}, \frac{622908012647232}{5},$$

$$\frac{77602878444025201997703040704}{5},$$

$$\frac{1204441348559630271252918141028336694332989128001036771264}{5}, \dots$$

(see A117596), which finally reaches an integer, a 57,735-digit number, after 18 steps!

If the fraction that we start with has denominator 2, we can say exactly how many steps are needed. If we start with $\frac{2l+1}{2}$ then we reach an integer in $m+1$ steps, where 2^m is the highest power of 2 that divides l. For example, $\frac{17}{2}$ (where $l = 2^3$) reaches the integer $1,204,154,941,925,628$ in four steps.

But even for denominator 3, we cannot say exactly what will happen. The following table shows what happens for the first few starting values. It gives the initial term, the number of steps to reach an integer, and the integer that is reached.

start :	$\frac{3}{3}$	$\frac{4}{3}$	$\frac{5}{3}$	$\frac{6}{3}$	$\frac{7}{3}$	$\frac{8}{3}$	$\frac{9}{3}$	$\frac{10}{3}$	$\frac{11}{3}$	$\cdots$
steps :	0	2	6	0	1	1	0	5	2	$\cdots$
reaches :	1	8	1484710602474311520	2	7	8	3	1484710602474311520	220	$\cdots$

(The second and third rows are sequences A072340 and A085276—they certainly stagger.)

Starting values of the form $\frac{l+1}{l}$ seem to take an especially long time to reach an integer. The examples $l = 5$ and 7 have already been mentioned. It is amusing to note that if we start with $\frac{200}{199}$ and repeatedly apply the approximately squaring operation, the first integer reached is roughly

$$200^{2^{1444}},$$

a number with about 10^{435} digits.

"If a Power Series Was a Power of a Power Series, What Power Would It Be, Seriously?"

That was the title of Nadia Heninger's talk about her 2005 summer project at AT&T Labs. Consider the following question, a typical problem from number theory. How many ways are there to write a given number n as the sum of four squares? That is, how many integer solutions (i, j, k, l) are there to the equation

$$n = i^2 + j^2 + k^2 + l^2?$$

Call the answer $r_4(n)$. Solutions with i, j, k, l in a different order or with different signs count as different, so for instance $r_4(4) = 24$, since for $n = 4$ (i, j, k, l) can be any of

$$(\pm 2, 0, 0, 0), (0, \pm 2, 0, 0), (0, 0, \pm 2, 0), (0, 0, 0, \pm 2), (\pm 1, \pm 1, \pm 1, \pm 1).$$

We can capture this problem in a generating function that looks like

$$R(x) := r_4(0) + r_4(1)x + r_4(2)x^2 + r_4(3)x^3 + \cdots,$$

in which the coefficient of x^n gives the answer $r_4(n)$. For this problem, it is easy to see that $R(x)$ is equal to the fourth power of Jacobi's famous "theta series"

$$\theta_3(x) := 1 + 2x + 2x^4 + 2x^9 + 2x^{16} + 2x^{25} + \cdots$$

(this is classical number theory: see for example Hardy and Wright [14] or Grosswald [11]). Of course this means that we can take the fourth root of $R(x)$ and still have integer coefficients.

What we (that is, Nadia Heninger, Eric Rains, and I [15]) discovered is that there are many other important generating functions for which it is possible to take a fourth root, or in general a kth root, and still have integer coefficients.

Many of our examples arise as "theta functions" of sphere packings. The most familiar sphere packing is the grocer's face-centered cubic lattice arrangement of oranges, which Tom Hales [13] has recently shown to be the densest possible sphere-packing of three-dimensional balls. For any lattice packing of balls in N-dimensional space, the theta series is a generating function whose coefficients give the number of balls with centers at a given distance from the origin. If there are M_d balls whose distance from the origin is $\sqrt{d}$, the theta series is

$$\sum_d M_d x^d. \tag{3}$$

The Jacobi theta series $\theta_3(x)$ mentioned above is simply the theta series of the one-dimensional lattice formed by the integers, and $R(x)$ is the theta series of the simple cubic lattice in four dimensions.

The starting point for our work was an observation of Michael Somos [26] that the 12th root of the theta series of a certain 24-dimensional lattice discovered by Gabriele Nebe ([23], also sequence A004046) appeared to have integer coefficients. We were able to establish his conjecture, and to generalize it to many other theta series and power series.

An example of one of our discoveries is this: the theta series of the densest lattice sphere packing in four dimensions is the fourth power of a generating function with integer coefficients. The theta series in question, that of the D_4 lattice packing, is

$$r_4(0) + r_4(2)x^2 + r_4(4)x^4 + \cdots \quad = \quad 1 + 24x^2 + 24x^4 + 96x^6 + 24x^8$$
$$+ 144x^{10} + \cdots, \tag{4}$$

and is formed by taking the even powers of x in $R(x)$. When we take the fourth root, we get

$$1 \quad + 6x^2 - 48x^4 + 672x^6 - 10686x^8 + 185472x^{10} - 3398304x^{12}$$
$$+ 64606080x^{14} - 1261584768x^{16} + 25141699590x^{18} - \cdots.$$

The coefficients stagger, changing sign at each step, and growing in size (A108092). Can a reader find any other interpretation of these coefficients?

In our paper, we begin by studying the more general question of when a power series of the form

$$F(x) = 1 + f_1 x + f_2 x^2 + f_3 x^3 + f_4 x^4 + \cdots$$

with integer coefficients is the kth power of another such power series. One of our results is that $F(x)$ is a kth power if and only if the series obtained by reducing the coefficients of $F(x)$ mod μ_k is a kth power, where μ_k is obtained by multiplying k by all the distinct primes that divide it. For example, to test if $F(x)$ is a fourth power, it is enough to check the series obtained by reducing the coefficients mod $\mu_4 = 4 \times 2 = 8$. Since all the coefficients of (4) except the constant term are divisible by 8, it reduces to 1 mod 8, and certainly 1 *is* a fourth power of a series with integer coefficients! So also is (4).

More dramatic examples are provided by the theta series of the E_8 lattice in 8 dimensions and the Leech lattice in 24 dimensions, which are respectively 8th and 24th powers of series with integral (albeit staggering) coefficients.

In [15] we give many other examples, including generating functions arising from weight enumerators of codes.

Dissections

My colleague Vinay Vaishampayan and I have found some surprising applications of the classical dissection problem to optical communications. But it seems that even the simplest questions in this subject are still unanswered. Since many Martin Gardner

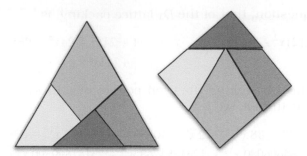

Figure 4. A triangle can be cut into four pieces that can be rearranged to form a square. Is it known that this cannot be done using only three pieces?

fans are experts in puzzles and their history, I mention this here, in the hope that some reader can provide further information.

The following is the simplest version of the question. It is known that any polygon can be cut up into a finite number of pieces that can be arranged, without overlapping, to form a square of the same area. (You are allowed to turn pieces over. To avoid complications caused by non-measurable sets, the edges of the pieces must be simple curves.) The question is, what is the minimal number, $d(n)$ say, of pieces that are required to dissect a regular polygon with n sides ($n \geq 3$) into a square? For the case $n = 3$, we are looking for a minimal dissection of an equilateral triangle to a square. There is a famous four-piece dissection, apparently first published by Dudeney in 1902 [6,8], shown in Figure 4. It seems unlikely that a three-piece dissection exists, but has anyone ever proved this? In other words, is $d(3)$ really 4?

As far as I know, none of the values of $d(n)$ are known for certain (except of course $d(4) = 1$). The best values presently known for $d(3), d(4), d(5), \ldots$ (A110312), taken from Frederickson [8] and Theobald [28], are

$$4?, 1, 6?, 5?, 7?, 5?, 9?, 7?, \ldots.$$

This is most unsatisfactory: the normal rule is that every term in a sequence in the OEIS should be known to be correct. This sequence is quite an exception, the values shown being merely upper bounds. It is not surprising that these entries stagger a bit—but are they correct?

The Kissing Number Problem

The N-dimensional kissing number problem asks for the maximal number of N-dimensional balls that can touch another ball of the same radius (the term comes from billiards). The problem has applications in geometry, number theory, group theory, and digital communications [5, 7]. For example, in two dimensions, six pennies is the maximal number that can touch another penny, as shown in Figure 5. This illustration shows a portion of the familiar hexagonal lattice packing in the plane. For the solution of the problem, however, the balls need not necessarily be part of a lattice packing. In dimensions 1 to 4, 8, and 24, the highest possible kissing number can be achieved using a lattice packing, but in dimension 9 there is a nonlattice packing with a maximal kissing number that is higher than is possible in any lattice packing, and this is almost certainly also true in ten dimensions—see [5]. If the record is achieved by a lattice packing, then we can read off the kissing number from the theta series defined in (3): this begins $1 + \tau x^u + \cdots$ for some u, where τ is the kissing number.

The answer in three dimensions is 12, and Musin [21, 22] has recently established the long-standing conjecture that the answer in four dimensions is 24, as found in the D_4 lattice packing—the number can be read off the theta series in (4). We also know the answers in dimensions eight and twenty-four (240 and 196560, respectively). The beginning of the sequence of solutions to the N-dimensional kissing number problem is

$$2, 6, 12, 24, 40?, 72?, 126?, 240, 306?, 500?, \ldots, 196560, \ldots$$

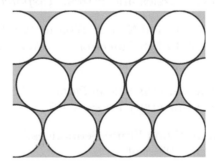

Figure 5. A portion of the hexagonal lattice packing of circles in the plane, illustrating the solution to the two-dimensional kissing number problem.

(see A001116). The entries with question marks are merely lower bounds. If an oracle offered to supply 64 terms of any sequence that I chose, I would pick this one. I would also ask the oracle for the constructions that it used—in particular, in high dimensions, are the arrangements always without structure, or is there an infinite sequence of dimensions where there are elegant algebraic constructions?

Bibliography

[1] W. Ackermann. "Zum Hilbertschen Aufbau der reellen Zahlen." *Math. Ann.* 99 (1928), 118–133.

[2] J. K. Aronson, quoted by D. R. Hofstadter in *Metamagical Themas* [17], p. 44.

[3] F. J. van der Bult, D. C. Gijswijt, J. P. Lindeman, N. J. A. Sloane, and A. R. Wilks. "A Slow-Growing Sequence Defined by an Unusual Recurrence." *J. Integer Seqs.* 10 (2007), #07.1.2.

[4] B. Cloitre, N. J. A. Sloane, and M. J. Vandermast. "Numerical Analogues of Aronson's Sequence." *J. Integer Seqs.* 6 (2003), #03.2.2.

[5] J. H. Conway and N. J. A. Sloane. *Sphere-Packings, Lattices and Groups*, Third Edition. New York: Springer-Verlag, 1998.

[6] H. E. Dudeney. "Puzzles and Prizes." *Dispatch*, May 4, 1902.

[7] Y. Edel, E. M. Rains and N. J. A. Sloane. "On Kissing Numbers in Dimensions 32 to 128." *Electronic J. Combinatorics* 5 (1998), #R22.

[8] G. N. Frederickson. *Dissections Plane and Fancy*. Cambridge, UK: Cambridge University Press, 1997.

[9] H. M. Friedman. "Long Finite Sequences." *J. Combin. Theory*, Ser A, 95 (2001), 102–144.

[10] S. W. Golomb. "Problem 5407." *Amer. Math. Monthly*, 73 (1966), 674; 74 (1967), 740–743.

[11] E. Grosswald. *Representations of Integers as Sums of Squares.* New York: Springer-Verlag, 1985.

[12] R. K. Guy. *Unsolved Problems in Number Theory*, Second Edition. New York: Springer-Verlag, 1994.

[13] T. C. Hales. "A Proof of the Kepler Conjecture." *Ann. of Math.* 162 (2005), 1065–1185.

[14] G. H. Hardy and E. M. Wright. *An Introduction to the Theory of Numbers*, Fith Edition. Oxford, UK: Oxford University. Press, 1979.

[15] N. Heninger, E. M. Rains, and N. J. A. Sloane. "On the Integrality of nth Roots of Generating Functions." *J. Combinatorial Theory*, Series A, 113 (2006), 1732–1745.

[16] Piotr Hofman and Marcin Pilipczuk. "A Few New Facts about the EKG Sequence." Preprint, 2007.

[17] D. R. Hofstadter. *Metamagical Themas.* New York: Basic Books, 1985.

[18] J. C. Lagarias. "The $3x+1$ Problem and Its Generalizations." *Amer. Math. Monthly* 92 (1985), 3–23.

[19] J. C. Lagarias, E. M. Rains, and N. J. A. Sloane. "The EKG Sequence." *Exper. Math.* 11 (2002), 437–446.

[20] J. C. Lagarias and N. J. A. Sloane. "Approximate Squaring." *Exper. Math.* 13 (2004), 113–128.

[21] O. R. Musin. "The Problem of the Twenty-Five Spheres" [in Russian]. *Uspekhi Mat. Nauk* 58:4 (2003), 153–154; English translation in *Russian Math. Surveys* 58 (2003), 794–795.

[22] O. R. Musin. "The Kissing Number in Four Dimensions." Preprint, 2005 (arXiv: math. MG/0309430).

[23] G. Nebe. "Some Cyclo-quaternionic Lattices." *J. Algebra* 199 (1998), 472–498.

[24] M. Sharir and P. K. Agarwal. *Davenport-Schinzel Sequences and Their Geometric Applications.* Cambridge, UK: Cambridge University Press, 1995.

[25] N. J. A. Sloane. *The On-Line Encyclopedia of Integer Sequences.* Available at http://www.research.att.com/~njas/ sequences/, 1996–2008.

[26] M. Somos. Personal communication, June, 2005.

[27] J. S. Tanton. "A Collection of Research Problems," *The Math Circle.* Available at http://www.themathcircle.org/ researchproblems, 2007.

[28] G. Theobald. "Square to Polygon Dissections." Available at home.btconnect.com/GavinTheobald/HTML/Square.html, 2006.

[29] D. Zagier. "The First 50 Million Prime Numbers." *Math. Intelligencer* 0 (1977), 7–19.

Seven Water Lilies

Péter Gábor Szabó and Zsófia Ruttkay

Figure 1. Seven water lilies. (Graphics designed by Ineke Lambers.)

In a square pool there are seven water lilies (see Figure 1), floating and growing, as long as there is enough free water surface. What final arrangements are possible, once the lilies have filled the pool?

We became interested in this problem in connection with a particular sangaku problem. *Sangaku* means "mathematical tablet" in Japanese. Sangakus were hung under the roofs of Shinto shrines and Buddhist temples from the mid-seventeenth century. Both the authors and the topics are varied. Geometrical problems, involving circles, were often introduced on a sangaku, in the form of concise and visually appealing illustrations [1].

The original sangaku invited the viewer to derive the radii of five congruent circles packed in a square. The second author was struck by the beauty of the visual arrangement in the geometrical problem, which suggested to her the subtle movement of water lilies on the lakes around her home and inspired her to see in the static drawing the final snapshot of a dynamical growth process. This view of the original, innocent problem raised a series of questions, leading back and forth in history, both into the deep waters of mathematics and into the domain of aesthetics.

The Beauty of Water Lilies

Initially we merely wanted to produce aesthetically pleasing arrangements. The mathematics discussed later just happened to emerge. We experimented with different numbers of lilies and also with lilies growing in a nonuniform, competitive way. This led us into the realm of beauty, which is difficult to model in mathematical terms. However, several arrangements evoked "ohs" from viewers.

Simulating the Growth Process

In order to simulate the growth process of water lilies of circular shape, we made a discrete computational model that is also available as a Java applet [3]. This allowed us to experiment with different arrangements of seeds, searching for arrangements that produce identical end results. We were surprised to learn that simulations similar to ours, but based on the physics of billiard balls to guide the movement of the circles, have been used to gain insight into the packing problem for large numbers of circles [4]. Since such numerical simulations do not provide exact results, it is necessary to prove that an arrangement suggested by a drawing actually exists, i.e., that the circles do not overlap.

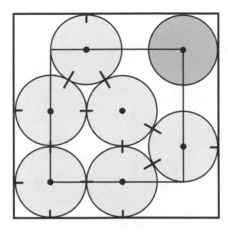

Figure 2. The densest packing of $n = 7$ equal circles in a square.

Dense Packing of Circles

In what arrangements can seven equal circles of maximum size be placed in a square without overlap? Figure 2 shows the proven optimal solution. Strangely there is only a computer-aided published proof for this result. This problem is the $n = 7$ instance of the famous circle packing problem—searching for the densest packing of n congruent circles in a square. In recent decades many interesting approaches have been devised to solve this circle packing problem, using a variety of algorithms, with the result that we now know the optimal solutions up to $n = 30$ circles. The hunt for optimal packings for $n > 30$ is still continuing [4].

Balls instead of Circles

If we step into three dimensions, we are faced with the problem of optimal packing of spheres. This was already an important practical problem in the 1600s, e.g., packing cannonballs in the densest possible way. Kepler conjectured that the method still in use today, e.g., in markets for making heaps of oranges, provides the optimal arrangement [5]. However, we had to wait almost 400 years before a proof was given. And it was done by checking a lot of possible cases a computer program. The correctness of this solution is still not fully accepted by expert mathematicians [6]. But what about

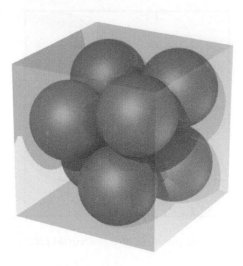

Figure 3. The densest known packing of $n = 7$ spheres in a cube, designed by Hugo Pfoertner [2].

the densest-seven-spheres-in-a-cube problem? Only a conjectured solution is known, which is shown in Figure 3.

Bibliography

[1] H. Fukagawa and D. Pedoe. *Japanese Temple Geometry Problems*. Winnipeg, Canada: The Charles Babbage Research Center, 1989.

[2] H. Pfoertner. "The Densest Packing of Equal Spheres in a Cube." Available at http://www.randomwalk.de/sphere/incube, 2005.

[3] Zs. Ruttkay. "Moving Sangakus: Growing Water Lilies." Available at http://wwwhome.cs.utwente.nl/~zsofi/sangaku/SanSim.html, 2006.

[4] P. G. Szabó, M. Cs. Markót, T. Csendes, E. Specht, L. G. Casado, and I. Garcia. *New Approaches to Circle Packing in a Square: With Program Codes*, Springer Optimization and Its Applications 6. New York: Springer, 2007.

[5] G. G. Szpiro. *Kepler's Conjecture: How Some of the Greatest Minds in History Helped Solve One of the Oldest Math Problems in the World.* Hoboken, NJ: John Wiley & Sons, 2003.

[6] "The Flyspeck Project." Available at http://code.google.com/p/flyspeck, 2008.

Part III

Puzzles and Games

Plate I. (See page 5.) The Pigs-in-Clover puzzle.

Plate II. (See page 5.) Judge's political Pigs-in-Clover cartoon.

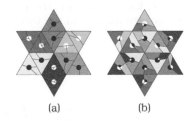

Plate III. (See page 11.) *PUCK* magazine cartoon about the 1880 presidential election.

Plate IV. (See page 243.) (a) Quarks (packaged). (b) Quarks (pied).

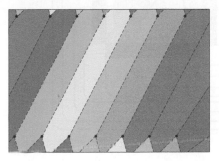

Plate V. (See page 29.) A dual map to K_7 on the torus (left), and a seven-colored map on the torus (right).

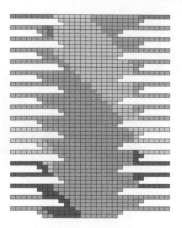

Plate VI. (See page 30.) The crochet pattern for 3/7 of the torus.

Plate VII. (See page 31.) The completed tori, positioned dually.

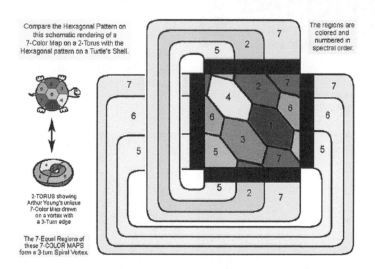

Compare the Hexagonal Pattern on this schematic rendering of a 7-Color Map on a 2-Torus with the Hexagonal pattern on a Turtle's Shell.

The regions are colored and numbered in spectral order.

2-TORUS showing Arthur Young's unique 7-Color Map drawn on a vortex with a 3-Turn edge

The 7-Equal Regions of these 7-COLOR MAPS form a 3-turn Spiral Vortex.

Plate VIII. (See page 81.) The seven-color map on the torus.

Plate IX. (See page 165.) "Elephant Spin Out" by Binary Arts.

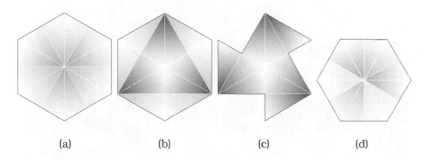

(a)   (b)   (c)   (d)

Plate X. (See pages 261 and 264.) (a) The Radial-2 position of the hexa-dodeca-flexagon always displays triangles from two different basic faces in this pattern. (b) The Alternate Hexagon position of the hexa-dodeca-flexagon can be composed of 12 triangles from one or two basic faces. (c) The Propeller of the hexa-dodeca-flexagon has only nine triangles and can be made from triangles from one or two basic faces. (d) The Radial-3 position appears when the hexa-dodeca-flexagon is flexed along two creases that run from the corners of the hexagon and two that run from midpoints on its sides. Triangles from three basic faces appear in a symmetrical pattern.

Plate XI. (See page 156.) A plane-filling f-tiling generated from a prototile based on an octomino. The f-tiling was carried through six generations, colored in the order red, orange, yellow, green, blue, and indigo in the center f-tiling.

Developing the Transmission Puzzle

M. Oskar van Deventer

This article tells the story behind the Transmission puzzle (see Figure 1). This puzzle has a board with four pins and four gears. The object is to put the gears on the pins and make them spin freely. When solving the puzzle, you will discover that you can put any three gears anywhere and they will always spin. However, when the fourth is added, they get stuck and don't spin anymore. So how does this puzzle work and why do the gears get stuck?

The best way to answer these questions is to describe the steps in the development of Transmission:

1. understanding involute gears,

2. equal-size unequal gears,

3. mathematical modeling,

4. many prototypes.

The first step was understanding involute gears. Involute gears are widely used because of their effective transmission of force.

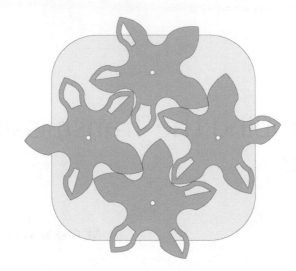

Figure 1. The Transmission puzzle.

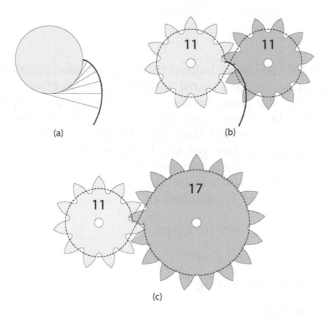

Figure 2. (a) Involute of circle. (b) Involute gears. (c) Unequal involute gears.

Figure 3. Unequal gears with identical diameters.

Figure 4. Gears with different gear sections.

George Miller gave me a reference to Gearology [1], which explains how to construct them. The shape of the teeth of these gears is the involute of a circle. This shape is made by winding a thin piece of wire around a pole, putting a pen at the end of the wire, and then unwinding the wire (see Figure 2(a)). The teeth are made by mirroring, rotating, and duplicating the involute shape. Finally, a gear is completed by making cycloid undercuts such that the tips of the teeth of one gear fit well between the teeth of another gear (see Figure 2(b)). Unequal gears are made by drawing the involutes of two unequal circles (see Figure 2(c)). As all of the teeth have the same size, the diameter of a gear is proportional to its number of teeth.

The second step was the "eureka moment"—realizing that that last remark is incorrect. By cheating a little bit, it is possible to construct gears with identical diameters but unequal tooth numbers (see Figure 3). There is a perfect fit between the two 12-tooth gears and also between an 11-tooth and a 12-tooth gear. There is a little play between the two 11-tooth gears. By combining sections from both types of gears, it is possible to construct gears that have non-constant gearing ratios (see Figure 4). When one gear is spun with constant speed, the other gear decelerates or acceler-

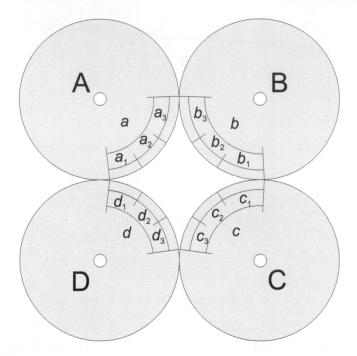

Figure 5. Equation $a_x + c_x = b_x + d_x$ makes the gears run smoothly.

ates. This explains why the four gears of the Transmission puzzle can get stuck, namely, when the net acceleration and deceleration in the loop of four gears is unequal to one.

The third step was the mathematical modeling of the puzzle. During an email discussion about the puzzle, George Miller found the equation that must be satisfied to have the gear run smoothly. If you put corresponding marks on each of the gears, as in Figure 5, it becomes quickly apparent that the angles of the segments a, b, c, and d satisfy the equation $a + c = b + d$. Consequently, this equation should be true for each of the subsegments as well: $a_1 + c_1 = b_1 + d_1$, $a_2 + c_2 = b_2 + d_2$, and $a_3 + c_3 = b_3 + d_3$. Armed with George's equation, Peter Knoppers wrote a solver program that checks all possible combinations of a specific set of gears and enumerates the solutions.

The fourth step was to make many different designs and prototypes to get everything right. This is the 90% part of the well-known 10%/90% inspiration/perspiration ratio. For a couple of months,

Figure 6. Many Transmission prototypes.

I made a new design every week, which Peter Knoppers cut on his laser cutter (see Figure 6). We tried tooth numbers ranging from 32 to 11, 8, 7, 5, 4 to even 3. We discovered a very elegant design for a complete set of 3-tooth gears that has a unique solution. Alas, it proved to be impossible to have the 3-tooth gears grip into each other. It was the same for 4-tooth gears. The 7-, 8-, and 11-tooth gears posed severe tolerance problems. Ultimately, 5 proved to be the right number of teeth for this puzzle. I chose to have the 5-tooth gears with 10 gear flanks, of which 6 are "thick" and 4 "thin." By a semi-systematic computer-assisted search, Peter and I found a set of gears that has a unique solution. This is the gear set shown in Figure 1 and coded in Table 1. More problems still needed to be solved. Because the tolerances are so tight, we had to find the exactly right compensation for the approximately 0.1 mm thickness of the laser cut. Moreover, the laser-cut holes gave too much friction. Therefore we drilled the central holes in the gears

Gear	Code									
A	1	0	0	0	1	1	1	1	0	1
B	1	0	0	1	0	0	1	1	1	1
C	1	0	1	1	0	0	1	1	1	0
D	1	0	1	0	1	1	1	1	0	0

Table 1. Coded Transmission design of Figure 1; 1 = thick flank, 0 = thin flank.

with a power drill, resulting in a smooth spin. The finishing touch was a suggestion by Peter Knoppers. By adding marker holes in the teeth, one can easily distinguish the "thick" and the "thin" gear flanks.

Peter Knoppers [2] made me 120 samples of Transmission, most of which I exchanged in Helsinki at IPP2005. So far only Nick Baxter has reported finding the solution by systematic analysis. If you want to make this puzzle for yourself, you can make a photocopy of Figure 1, glue it onto sheet material, and cut out the shapes with a jigsaw. Notice that high accuracy is required. Alternatively, you can contact me for the electronic drawings at m.o.vandeventer@planet.nl.

Bibliography

[1] Boston Gear. "Gearology." Available at http://bostongear. com/training/gearology.asp, 2003.

[2] Peter Knoppers. *Buttonius Puzzles & Plastics*. Available at http://www.buttonius.com/.

Triple-7 Hamiltonian Chess

David S. Dillon, Jeremiah Farrell, and Tom Rodgers

The new puzzle game we have in mind involves seven each of the three minor pieces in chess; the rook, the bishop, and the knight. It is played on the mutilated 5 × 5 board shown in Figure 1 with the

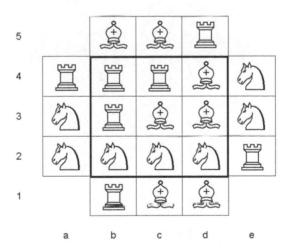

Figure 1. The 5 × 5 board.

pieces permanently marked as shown on the remaining squares after the four corners are deleted.

The Puzzle

Obtain 20 small tokens (pennies will do) and try to place all 20 on the board by using the following rule. Start in any open square, then make a move appropriate to the chess piece marked on the opening square to an unoccupied square and drop a token there. The moves may cross occupied squares but must start and end on unoccupied ones. Also, to avoid trivialities, we insist that no move from chess piece to like chess piece be allowed (i.e., no knight to knight, etc.). It is as if there were three armies—the knights, the rooks, and the bishops—each fighting the others' forces.

If you can place all 20 tokens you have solved the puzzle. There are thousands of ways of solving this puzzle (more on this later) but, remarkably, most people have great difficulty in finding even one.

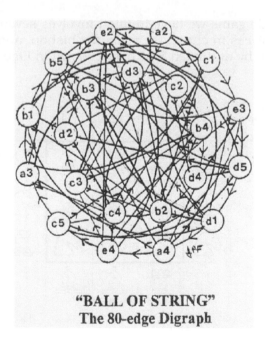

"BALL OF STRING"
The 80-edge Digraph

Figure 2. The full graph of moves.

The Game

Two persons alternately make moves as in the puzzle version and the last player able to make a legal move wins. This entertaining game can never end in a draw and both players, after some experience, have many opportunities for strategic play. Typically, several empty squares remain after the game ends. If there is an even number of empties the first player wins, otherwise the second. The full graph of moves to the puzzle game will have 80 directed edges on 21 nodes, and this alone explains the difficulty in finding solutions. A drawing of this complex diagram appears in Figure 2.

Many solutions will take the form of an elegant Hamiltonian circuit tour of the board. Named after the nineteenth-century Irish mathematician Sir William Rowan Hamilton, these tours visit every square of the board exactly once and return to the starting square

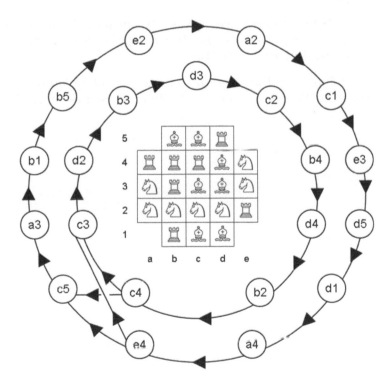

Figure 3. A solution to the puzzle game.

when the tour is completed. One of us (Dillon) performed a computer search and found 2,070 Hamiltonian circuits for our digraph. For more information on W. R. Hamilton and the modern "traveling salesman" problem, see Martin Gardner's article [2, Chapter 6].

As an introduction to the puzzle game, one can play "Triple-3s" on the inner 3×3 board. It is best to demand full Hamiltonian circuits in this version and then the solution is unique (of course, the empty square can be any of the nine squares).

A solution to the puzzle game appears in Figure 3. The inner circle is the unique tour of unlike pieces on the 33 board. The outer circle is one of eight tours of unlike pieces on the outer twelve squares. Specifically, a solution to the puzzle could start: c3 - d2 - b3 - d3 - c2 - b4 - d4 - b2 - c4 . . . .

Bibliography

[1] Jeremiah Farrell and Eric Nelson. "Games on Nonsymmetric Configurations." *Word Ways: The Journal of Recreational Linguistics* 38:2 (2005), 106–109.

[2] Martin Gardner. *The Scientific American Book of Mathematical Puzzles & Diversions.* New York: Simon & Schuster, 1959.

Retrolife

Yossi Elran

The "Game of Life" is one of the most popular games ever invented. The famous British mathematician, John Horton Conway, invented the game in the early 1970s [1]. Martin Gardner was so intrigued by it that he dedicated three consecutive chapters in his book *Wheels, Life and Other Mathematical Amusements* to it [2].

"Life" is essentially a one-player game played on an $n \times n$ board or grid. It can be played on an $n \times n$ checkerboard or, as is much more common nowadays, as a computer simulation. Initially, the player forms a shape by placing checkers on the checkerboard, or by coloring in cells on the board generated by the computer simulation. A cell that is colored (or a cell occupied by a checker) is called a "live" cell, while an empty cell is a "dead" cell. Every cell on the board has exactly eight neighboring cells: one in each direction (up, down, right, and left) and four on the diagonals. The shape then evolves in time, through a number of discrete time steps known as "generations," according to a simple set of rules:

1. A "dead" cell is "born" (i.e., the cell is colored) if it has exactly three live neighboring cells.

2. A "live" cell "dies" if it has less than two neighbors, or more than three neighbors.

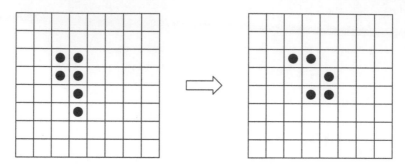

Figure 1. An example of the evolution of an initial shape in "Life" for one
generation.

3. A "live" cell "survives" (i.e., a cell that is colored remains so) if
   it has either two or three neighbors.

Figure 1 shows an example of how an initial generation evolves for
one time step.

Playing the game is very simple and straightforward. After the
initial shape is placed on the board, it evolves in time, producing a
very wide variety of outcomes. The name of the game, "Life," is due
to this evolutionary behavior. An important aspect of the game is
that applying the rules to any given generation results in only one
possible outcome. The reverse, however, is not true. Finding the
preceding generation for a given shape is a difficult task—in fact,
it is thought to be NP-complete, making it an interesting problem
to pose as a puzzle. In this puzzle, which we coin "Retrolife," the
task is to find predecessors for a given initial state in the "Game of
Life" that comply with some extra constraints. For clarity, we will
call the shape we want to construct (the initial state in the "Game
of Life") the "puzzle" and its predecessor(s) the "solution."

Theoretically, there is an infinite number of states in the preceding generation for any given shape, since—for any given solution—
one can always add a lone "live" cell (one that has no "live" neighbors) anywhere on the infinite "solution" board, and it will con-
veniently die when the rules of "Life" are applied correctly to give
the required "puzzle" board. In "Retrolife," we therefore demand
that the "solution" board contains no redundant "live" cells. This
means that there should be no "live" cells on the "solution" board
that do not contribute to the formation of the solution. "Retrolife"
becomes more interesting and attractive if we add additional con-

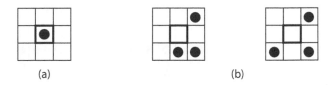

(a)   (b)

Figure 2. (a) The simplest "Retrolife" puzzle. (b) Two possible solutions using Rule 1.

straints. One such constraint, Rule 1, is that all the cells defining the required shape on the "puzzle" board have to be cells that are "born", i.e., they cannot "survive"—they cannot be "live" cells on the "solution" board.

There can be many solutions to a "Retrolife" problem; sometimes there is a unique solution and sometimes there is no solution. If there is no solution even when there are no extra constraints, the state is the well-known "Garden of Eden" in "Life". The very simplest "Retrolife" puzzle is a board containing just one live cell. The puzzle is, of course, to find the predecessor to this shape according to Rule 1, i.e., the cell itself has to be "born" when applying the rules of "Life" to the "solution" board. There are a number of solutions to this problem. Figure 2 shows two possible solutions.

Finding the solution (i.e., the predecessor) to a row of two "live" cells is also trivial. A solution for any linear shape according to Rule 1 can always be found. Finding solutions for other shapes, such as the shapes that form numbers or letters of the alphabet, proves to be a more complicated task and demonstrates some of the difficulties that one runs into when trying to solve "Retrolife" puzzles. Specifically, one might run into an annoying situation where a cell "survives" or is "born" somewhere on the board other than within the boundaries of the desired shape. In these cases, extra "live" cells must be used in order to get rid of the unwanted additions.

Figure 3 shows the puzzle and solutions for the number "7" and the letters "O" (a unique solution) and "M." There is no possible solution according to Rule 1 for the small-capped letter "g," or in fact for any letter that can be regarded as the letter "o" with a "tail." The joint between the "head" and the "tail" of the letter forces a violation of the rule.

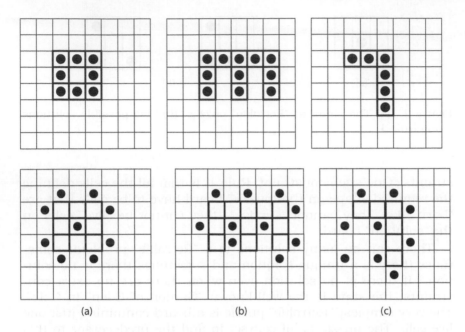

Figure 3. Puzzles (top) and possible solutions (bottom) to "Retrolife" puzzles (a) "O," (b) "M," and (c) "7."

There are, of course, many different rules that can be chosen as extra constraints for "Retrolife." Instead of demanding that all the cells that form the shape on the "puzzle" board be "born," we can demand that all of them "survive" (Rule 2) or that half "survive" and half are "born" (Rule 3). When the shape is constructed using an odd number of cells, q, we demand that $\frac{1}{2}(q-1)$ cells are "born" and $\frac{1}{2}(q+1)$ cells "survive."

As in the case of Rule 1, all linear shapes can be constructed using these rules, according to a straightforward procedure in each case. The small-capped "g" and similar letters cannot be constructed using either rule. The letters "O" and "M" and the number "7" can be constructed using Rule 3. It is impossible to construct the letters "O" and "M" according to Rule 2, since the "puzzle" shapes for these letters contain at least one "live" cell that has more than three neighbors. This cell cannot "survive" for more than one generation; hence there is no solution in this case. Fig-

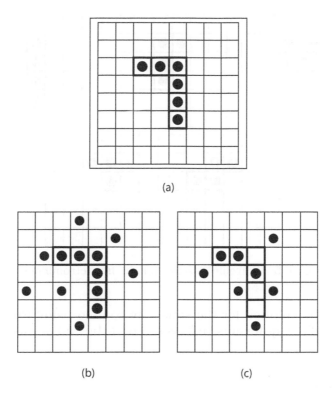

(a)

(b)   (c)

Figure 4. (a) The "Retrolife" puzzle for the number "7." (b) Possible solution according to Rule 2. (c) Possible solution according to Rule 3.

ure 4 presents possible solutions for the number "7" according to Rule 2 and Rule 3.

There is no end to the number of "Retrolife" puzzles that can be created. The puzzle can be refined in many ways: more complicated shapes can be studied, the constraints can be changed or modified, and the process can be continued for more than one generation backwards.

An interesting challenge is to insist that the solution is one with the minimum possible number of "live" cells. Figure 5 shows the "puzzle" boards for the five tetrominoes. Perhaps the reader would like to find the solutions for these shapes according to the three rules described above, using the smallest possible number of "live" cells in the solution. Most of the solutions are not that difficult;

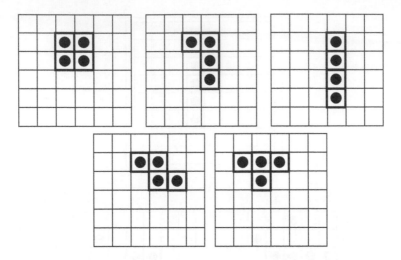

Figure 5. "Retrolife" puzzles for the five tetrominoes.

however, the solution for the square tetromino using Rule 1 with 14 "live" cells is quite elusive (it is given at the end of this article).

Solutions

The "Retrolife" solution for the square tetromino using Rule 1 with 14 "live" cells is illustrated in Figure 6.

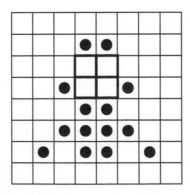

Figure 6. Solution to the square tetromino puzzle in Figure 5.

Tetromino	Rule 1	Rule 2	Rule 3
Square	14	4	4
Straight	6	8	6
L	5	7	5
Z	6	6	4
T	7	7	5

Table 1. Minimum number of "live" cells required for the solutions to the tetromino puzzles in Figure 5.

Table 1 shows the minimum number of "live" cells required for each solution. Special thanks is given to Dr. Oded Margalit from IBM Haifa Research Labs for writing a computer program to verify these results.

Bibliography

[1] Elwyn R. Berlekamp, John H. Conway, and Richard K. Guy *Winning Ways for Your Mathematical Plays*, Vol. 4. Wellesley, MA: A K Peters, Ltd., 2004.

[2] Martin Gardner. *Wheels, Life and Other Mathematical Amusements*. New York: W. H. Freeman, 1983.

The Logologicomathemagical 7 × 7 Puzzle

Jeremiah Farrell and Robert Friedhoffer

Here are the preliminaries to the puzzle:

(I) Toss a coin and remember how it lands (if on edge, toss it again).

(II) To find your lucky number, take your age and subtract from it the sum of its digits and then add 2. For example, if you are 103, you subtract 4 and add 2 for a lucky number of 101. To find your lucky letter, go to the MOUSETRAP wheel in Figure 1 and count your lucky number clockwise from the M until you land on your lucky letter. So if your lucky number was 7, your lucky letter would be R.

(III) Neil Abbott and Neal Costello are next to bat. Choose an order for them to bat.

(IV) Choose any three-digit number in which not all the digits are alike. Arrange the digits in descending order, reverse them to make a new number, and subtract the new number from the old. Call this number X. (Preserve zeroes if they occur.)

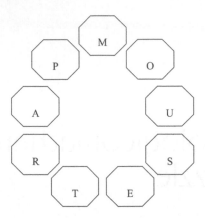

Figure 1. The MOUSETRAP wheel for II.

After a few repeats of this process you will reach a constant X. Decode this constant using A= 1, B= 2, ... and silently write the word here _____.

(V) You and I are in a room and one of us departs. Rocko shouts "Whooz left?"

Now we are ready for the puzzle:

1	2	3		4	5	6
7				8		
9			10			
		11				
12	13				14	15
16				17		
18				19		

Across

1. What transcendental and denumerable have in common

4. Has an anagram with the same meaning

7. Latin hail

8. Contraries

9. (From I) Your coin landed _____.

11. Usually found in books

12. Did seem irregular?

16. Word from Shakespeare

17. French checker?

18. 007's base of operations?

19. One-half of 12?

Down

1. Brooklynese emonstrative

2. Girl's name

3. (From III) They both draw walks. Therefore it *must* be true that _____ first.

4. (From II) The _____ is your lucky letter.

5. (From V) The answer can always be ___.

6. How we anticipated 9 across

10. Certain male

12. Could be a surface oval

13. After either ups or downs?

14. One, if used enough can make this

15. (From IV) The end word

Here are the answers:

Across

1. DEN (contained in both words)

4. AYE

7. AVE

8. NOS

9. _____UP, heads or tails fits[1]

11. LAW

12. MISDEED[2]

16. ADO

17. ROI

18. TEN

19. ONE[3]

Down

1. DAH or DAT

2. EVE or EVA

3. NE_LS ON, either I or A fits[4]

4. ANSWER O[5]

5. YOU[6]

6. ESP

10. LAD or DAD

12. MAT

13. IDE

14. EON

15. DIE[7]

Notes:

1. See [1], where Will Shortz, editor of the *New York Times* cross-word puzzle, explains Jeremiah Farrell's election day 1996 puzzle that predicted the Clinton–Bob Dole race. "As I've often said," Shortz remarks, "this is my favorite crossword of all time."

2. An apt anagram, cryptic style.

3. $12 = 1$ and 2.

4. One of them is "on (base) first" and one is "on first (base)."

5. This is one of many "casting out nines" variations.

6. You have either left the room or are left in the room. Rocko's accent is ambiguous.

7. The constant is always 495, which translates to DIE but remember "never say die." See [2].

Bibliography

[1] Coral Amende. *The Crossword Obsession.* New York: The Berkley Publishing Group, 2001.

[2] Martin Gardner. "Self-Numbers." In *The Colossal Book of Short Puzzles and Problems*, edited by Dana Richards, p. 70. New York: W. W. Norton, 2006.

Seven Roads to Roam: A Magical Journey

Jeremiah Farrell and Judith H. Morrel

Here we present the effect and the method for a mathematical magic trick that takes the audience on a journey...

The Effect

Our magic begins by the magician showing the subject the road-map wheel with the eight nodes labeled ASTEROID and the seven numbered routes between them. The wheel is shown in Figure 1.

"When my back is turned, choose a node at random without telling me your start," explains the magician. "And then take a journey on the roads calling out your route numbers as you go."

The subject travels as long as desired and, when finished, tells the magician "I have arrived."

The magician asks the subject "Without telling me whether it is your start node or your finish node, tell me one of the two."

The subject does so, and the magician immediately is able to name the other node of the journey.

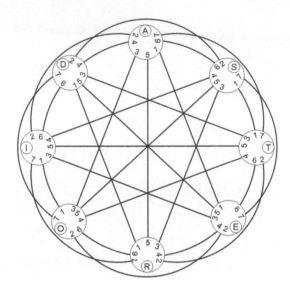

Figure 1. The wheel.

The Method

The wheel is actually a depiction of the complete graph K_8, and the roads are labeled so that the numbers 1 through 7 (along with an identity 0) follow the finite group $Z_2 \times Z_2 \times Z_2$ whose addition table appears in Table 1.

This is one of the five finite groups of order 8 and the one that seems especially difficult for lay persons to see through, but remains easy enough for "mathemagicians" to use. Many readers

	0	1	2	3	4	5	6	7
0	0	1	2	3	4	5	6	7
1	1	0	3	2	5	4	7	6
2	2	3	0	1	6	7	4	5
3	3	2	1	0	7	6	5	4
4	4	5	6	7	0	1	2	3
5	5	4	7	6	1	0	3	2
6	6	7	4	5	2	3	0	1
7	7	6	5	4	3	2	1	0

Table 1. The addition table for the finite group $Z_2 \times Z_2 \times Z_2$.

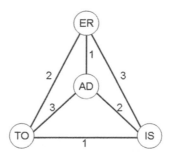

Figure 2. A smaller wheel.

will recognize this group table as nim sum addition or base-two addition without carrying. For example $3 + 6 = (2+1) + (4+2) = 011 + 110$ (Base 2) $= 101 = 5$. With a little practice, the magician is able to quickly perform each addition as the numbers are called out and is always able to reduce the journey to one number. If the roads called were 1, 3, 6, 5, and 4, the magician calculates $1 + 3 = 2$, $+6 = 4$, $+5 = 1$, $+4 = 5$, and knows that the start and end nodes are separated by road 5. Thus starting on, say T, the journey travels to I and given either I or T the magician names the other node.

The smaller graph in Figure 2 is K_4, and the road numbers use 1, 2, and 3 as labels. Adding 0 gives the group action called the Klein 4-group whose table is the 4×4 upper left corner of our nim sum table. The node labels AD, ER, IS, and TO are designed as a mnemonic for our eight-node graph. The four vowels are separated by roads 1, 2, and 3 on the K_8 graph as are the four consonants. Road 4 connects the pairs A-D, E-R, I-S, and T-O. To find other connections, say A to T, travel in the K_4 from A to D (a 4) and from D to T (a 3) for a total $4 + 3 = 7$ as the route from A to T.

Fractal Tilings Based on Dissections of Polyominoes, Polyhexes, and Polyiamonds

Robert W. Fathauer

Fractal tilings ("f-tilings") are described based on single prototiles derived from dissections of polyominoes, polyhexes, and polyiamonds. These prototiles have one or two long edges and two or more short edges, and the angles between long and short edges are in most cases irrational. The f-tilings are constructed by iterative arrangement, according to a simple matching rule, of successively smaller generations of tiles about a central group of largest-generation tiles that form the generating polyomino, polyhex, or polyiamond. For the most part, these f-tilings do not cover the infinite plane, but rather are bounded and contain singular points. Within their boundaries, which are in most cases fractal curves, they contain neither gaps nor overlaps, and the f tilings presented here are all edge-to-edge or pseudo-edge-to-edge.

Fractals and tilings can be combined to form a variety of esthetically appealing constructs that possess fractal character and at the same time obey many of the properties of tilings. Previously, we described families of fractal tilings based on kite- and dart-shaped quadrilateral prototiles [4], v-shaped prototiles [5], and prototiles constructed from segments of regular polygons [3]. Many of these constructs may be viewed online [2]. These papers appear to be the first attempts at a systematic treatment of this topic, though isolated examples were earlier demonstrated by M.C. Escher [1] and Peter Raedschelders [8].

In Grünbaum and Shephard's book *Tilings and Patterns* [7], a *tiling* is defined as a countable family of closed sets (tiles) that cover the plane without gaps or overlaps. The constructs described in this paper do not for the most part cover the entire Euclidean plane; however, they do obey the restrictions on gaps and overlaps. To avoid confusion with the standard definition of a tiling, these constructs will be referred to as "f-tilings," for fractal tilings.

The tiles used here are "well behaved" by the criterion of Grünbaum and Shephard; namely, each tile is a (closed) topological disk. Most of the f-tilings explored in [2–5] are edge-to-edge; i.e., the corners and edges of the tiles coincide with the vertices and edges of the tilings. However, they are not "well behaved" by the criteria of normal tilings; namely, they contain singular points, defined as follows. Every circular disk, however small, centered at a singular point, meets an infinite number of tiles. Since any f-tiling of the general sort described here will contain singular points, we will not consider singular points as a property that prevents an f-tiling from being described as "well behaved." These f-tilings provide a rich source of unique fractal images and also possess considerable recreational mathematics content.

The prototiles considered in this paper are derived by dissecting polyominoes, polyhexes, and polyiamonds. These are shapes made by connecting squares, regular hexagons, and equilateral triangles, respectively, in edge-to-edge fashion. Polyominoes made from n squares will be called n-ominoes, and similarly for polyhexes and polyiamonds, except for polyominoes made up of a small number of squares, which will be referred to using standard prefixes such as "hex" for 6. For a discussion of different types of polyominoes, polyhexes, and polyiamonds, and conventional tilings using them, see [7].

In this paper, we construct f-tilings from prototiles created by dissecting these shapes, as illustrated in Figure 1. A 24-iamond is shown at upper left in this figure, dissected into six congruent tiles. Note that each tile has two long edges and four short

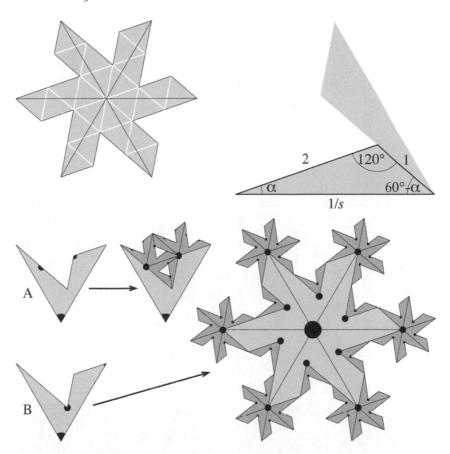

Figure 1. Derivation of a prototile and its application to the construction of an f-tiling.

edges, where the short edges are defined by the edges of the constituent triangles in the polyiamond. (A distinction between true edges and pseudo-edges is made below.) An f-tiling is created by fitting smaller tiles around larger tiles, where a long edge of the smaller tile has the same length as a short edge of the larger tile. At upper right in Figure 1, the scaling factor s and angle of rotation α between successive generations of tiles are indicated. Straightforward application of trigonometry reveals that α and s are both irrational, with approximate values of 19.11° and 0.3780. At lower left, two possible matching rules for arranging the smaller tiles are

Figure 2. Construction of an f-tiling using the prototile from Figure 1 along with Matching Rule B, carried through the first six iterations.

marked with black dots. Matching Rule A does not allow tiling without overlaps, while Matching Rule B does. The first two iterations of the construction of the f-tiling using Matching Rule B are shown at lower right.

In order to generate the full f-tiling, the same matching rule is applied repeatedly with successively smaller generations of tiles. In theory, an infinite number of such iterations should be performed. In practice, the overall structure of the f-tiling changes little after roughly a half dozen iterations due to the fact that the tiles shrink rapidly. This simple process generates f-tilings with fractal boundaries that can be quite complex. Figure 2 illustrates this phenomenon by continuing the construction begun in Figure 1. A portion of the f-tiling is shown after each of the first six iterations. In this particular f-tiling, an intricate channel forms. It is clear from the similarity of the channel from one iteration to the next that this channel will become infinitesimally narrow with repeated iteration, but never fully close off.

In general, the boundaries of f-tilings are fractal curves, though there are cases in which the boundaries are non-fractal polygons. The boundaries are similar to Koch islands and related constructs, in which a line segment is distorted into multiple smaller line segments, which are in turn distorted according to the same rule, and so on.

Now that the overall process is a little clearer, we can examine possible prototiles in more depth. Due to space limitations, only polyomino prototiles will be described in detail. The major features of polyhex and polyiamond prototiles will be summarized after the polyomino discussion.

Most polyominoes have adjacent straight-line segments longer than the edges of the constituent squares, as shown in Figure 3. For this reason, there are relatively few true edge-to-edge f-tilings using prototiles created by dissecting polyominoes. In order to provide a richer variety of examples, pseudo-edge-to-edge f-tilings will be considered as well. In these cases, the edges of the constituent squares are considered to define pseudo-edges of the polyominoes, as shown in Figure 3.

Each prototile has one or two long edges and two or more short edges. The angles between the long edges and adjacent short edges are for the most part irrational, but sum to a multiple of $\pi/2$, as shown in Figure 3. The f-tilings are constructed by first matching the long edges of identical prototiles, such that the group of

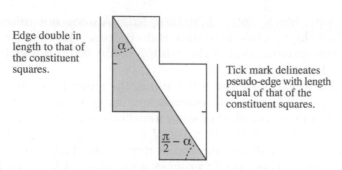

Edge double in length to that of the constituent squares.

α

Tick mark delineates pseudo-edge with length equal of that of the constituent squares.

$\frac{\pi}{2} - \alpha$

Figure 3. A tetromino with tick marks indicating pseudo-edges and irrational angle α.

first-generation tiles forms the dissected polyomino. Other more complex arrangements of first-generation tiles are possible. In addition, f-tilings may be constructed that possess more than one prototile, but these will not be considered here.

Three requirements simplify the search for polyomino-based prototiles that allow f-tilings.

1. *The generating polyomino must have two-fold or four-fold rotational symmetry.* While polyominoes without rotational symmetry may be used, they generate no new prototiles, and therefore they do not need to be considered. While not proven here, this is made readily apparent by examining a table showing all polyominoes possible for a given number of squares. In addition, mirrored variants of polyominoes are not considered to generate distinct prototiles, as they would result in f-tilings that are an overall mirror of the f-tilings constructed from non-mirrored variants.

2. *For a prototile generated by bisecting a polyomino, each bisecting line, which will form the long edge of the prototile, must originate and terminate at corners of the polyomino and pass through the centroid of the polyomino.* If the endpoints aren't at corners or pseudo-corners, the short edges will not all be of the same length. The long edges of the prototile must be longer than the short edges or pseudo-edges of the prototile.

3. *For a prototile generated by dissecting a polyomino into four equal parts, each dissecting line, which will form one long edge*

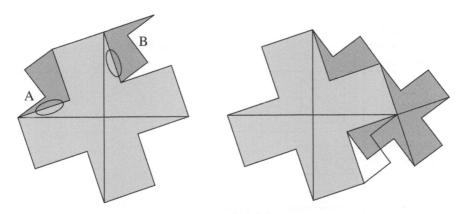

Figure 4. For prototiles generated from four-fold polyominoes, there are two choices, A and B, for arranging smaller tiles around large tiles. In order to avoid overlaps using a consistent matching rule, four-fold prototiles must have an even number of short edges.

of the prototile, must run from the centroid of the polyomino to a corner. The four dissecting lines are related to each other by rotations of 90° about the centroid. Again, the long edges of the prototile must be longer than the short edges or pseudo-edges of the prototile. For reasons shown below, the number of short edges or pseudo-edges must be even. This rules out any polyomino made up of $4n + 1$ squares.[1] The only other polyominoes with four-fold symmetry are made up of $4n$ squares, so only these need be considered.

Figure 5 shows candidate polyominoes made up of one to five squares and prototiles that meet these criteria. Prototiles colored gray allow f-tilings, while those colored black do not. Only gray prototiles without tick marks allow true edge-to-edge f-tilings.

In the case of prototiles derived from polyominoes with four-fold rotational symmetry, there are two choices of matching rules for arranging tiles of a given generation around tiles of the next larger

---

[1]The only pentomino with four-fold rotational symmetry yields prototiles with three short edges. Adding four squares to this pentomino yields prototiles with either three or five short pseudo-edges. It can easily be seen that adding four squares to any four-fold polyomino will either add two short pseudo-edges, leave the number of short pseudo-edges unchanged, or subtract two short pseudo-edges. The number of short pseudo-edges for prototiles is therefore always odd, which does not allow tiling, as demonstrated in Figure 4.

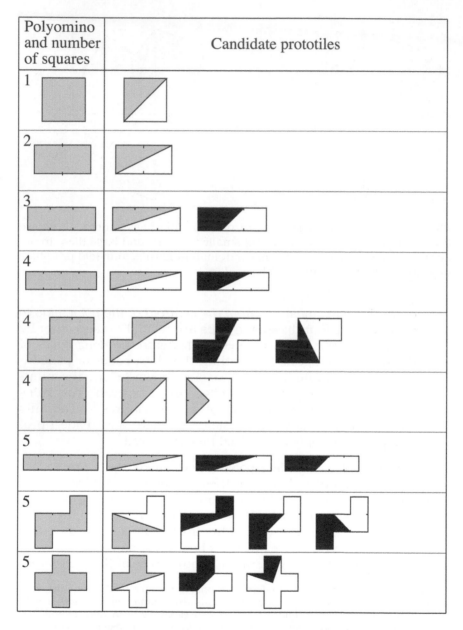

Figure 5. Candidate prototiles for polyominoes made up of one to five squares. The gray ones tile, while the black ones do not.

generation. This is illustrated in the left side of Figure 4, where the two choices are labeled A and B. The right side of Figure 4 shows why the number of short edges cannot be odd if a single matching rule is used. The fact that there are two long edges for each smaller tile requires an even number of short edges for each larger tile if overlaps are to be avoided.

In the case of polyhexes, two adjacent edges cannot lie in a straight line, so all f-tilings formed are true edge-to-edge f-tilings. Only polyhexes with two- and three-fold rotational symmetry need be examined. It can be shown that six-fold f-tilings based on poly-hex prototiles will not work [6]. The problem is analogous to that illustrated in Figure 4 for polyominoes; namely, the prototiles always have an odd number of short edges. A table similar to that of Figure 5, but for polyhexes, is found in [6].

In the case of polyiamonds, most candidate prototiles will result in pseudo-edge-to-edge f-tilings, such as those in Figures 1 and 2. For polyiamonds, two-, three-, and six-fold f-tilings are allowed.

Now that the prototiles have been described in greater detail, we give a number of examples of f-tilings constructed from these prototiles. Our first examples have overall two-fold rotational symmetry. The starting point for an f-tiling of this sort is a pair of tiles that form the generating polyomino, polyhex, or polyiamond. These are surrounded by smaller tiles in (pseudo-) edge-to-edge fashion, with the construction process proceeding iteratively. The only option is whether or not the tiles are mirrored between successive generations, but no mirrored variants are considered here due to space limitations.

In Figure 6, we show two examples of f-tilings of this sort. The left f-tiling is based on the "X" pentomino and is edge-to-edge, with a reduction factor of $1/\sqrt{10}$ between tiles of successive generations. This number is easily obtained by noting that the diagonal is that of three squares in a row and then applying the Pythagorean theorem. (If the squares have edges of length 1, the long edge of the prototile has length $\sqrt{1^2 + 3^2}$.) Between successive generations the tiles are rotated by $\arctan(1/3) \approx 18.43°$ (plus multiples of $\pi/2$ as required for fitting a particular edge). The right f-tiling in Figure 6 is based on a hexiamond and is pseudo-edge-to-edge, with a reduction factor of $1/\sqrt{7}$.

In Figure 7, we show two f-tilings constructed from the same dissection of a 4-hex, using two different matching rules for mating

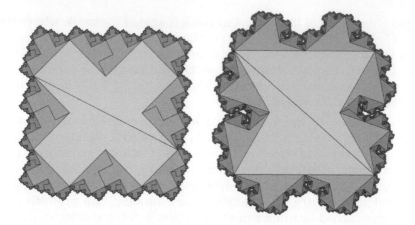

Figure 6. (Left) A true edge-to-edge f-tiling generated from a prototile based on the "X" pentomino. (Right) A pseudo-edge-to-edge f-tiling generated from a prototile based on a hexiamond.

Figure 7. Two f-tilings generated from a prototile based on a 4-hex, in which the scaling factor between successive generations is approximately 0.3780. Different matching rules are used for the two.

smaller tiles to larger tiles. Both f-tilings have three-fold rotational symmetry and complex fractal boundaries, but the shapes of the boundaries are quite different.

For some f-tilings, such as the one shown at left in Figure 7, holes form that just fill in the infinite limit. From a recreational

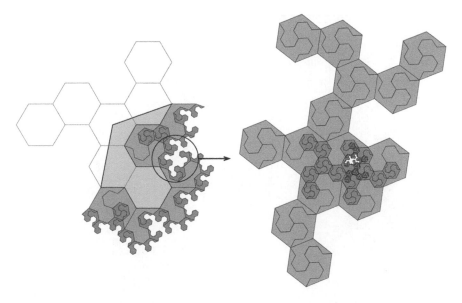

Figure 8. An example of a 14-hex-shaped hole that persists from generation to generation in an f-tiling based on a dissection of a 7-hex.

mathematics standpoint, these holes can be quite interesting. They can be examined by starting with a large polyhex and tiling inward. An example of a relatively complex polyhex-shaped hole (a 14-hex) that persists from generation to generation, though rotated, is shown in Figure 8. The generating 7-hex is shown at left, along with a portion of the f-tiling, which illustrates the origin of the hole. At right, the hole is shown filling in through four generations of tiles.

In some cases, an entire f-tiling constitutes a supertile that tiles the plane in conventional fashion. A four-fold example based on a prototile created by dissecting an octomino is shown in Figure 9. A six-fold example based on a prototile created by dissecting a 36-iamond is shown in Figure 10.

We have presented several examples of fractal tilings (f-tilings) based on prototiles derived by dissecting polyominoes, polyhexes, and polyiamonds. There are an infinite number of f-tilings of this sort. However, in general they become increasingly less interesting for higher-order polyominoes, polyhexes, and polyiamonds, due to the fact that the scaling factor between successive generations

Figure 9. A plane-filling f-tiling generated from a prototile based on an octomino. The f-tiling was carried through six generations, with each generation a darker shade of gray in the center f-tiling. (See Color Plate XI.)

becomes more extreme. Fractal tilings that result in recurring polyhex-shaped holes have also been demonstrated, as well as f-tilings that form supertiles that in turn tile the infinite mathematical plane.

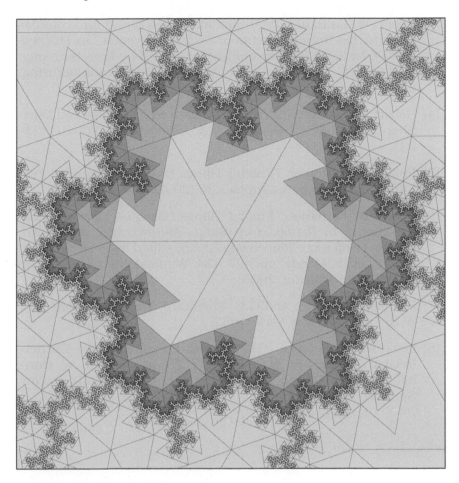

Figure 10. A plane-filling f-tiling generated from a prototile based on a 36-iamond. The f-tiling was carried through six generations, with each generation a darker shade of gray in the center f-tiling.

Bibliography

[1] Bruno Ernst. *The Magic Mirror of M.C. Escher.* New York: Ballantine Books, 1976.

[2] Robert W. Fathauer. "Compendium of Fractal Tilings." Available at http://www.mathartfun.com/shopsite_sc/store/html/Compendium/encyclopedia.html, 2000/2008.

[3] Robert W. Fathauer. "Self-Similar Tilings Based on Prototiles Constructed from Segments of Regular Polygons." In *Proceedings of the 2000 Bridges Conference*, edited by Reza Sarhangi, pp. 285–292. Winfield, KS: Central Plain Book Manufacturing, 2000.

[4] Robert W. Fathauer. "Fractal Tilings Based on Kite- and Dart-Shaped Prototiles." *Computers & Graphics* 25 (2001), 323–331.

[5] Robert W. Fathauer. "Fractal Tilings Based on v-shaped Prototiles." *Computers & Graphics* 26 (2002), 635–643.

[6] Robert W. Fathauer. "Fractal Tilings Based on Dissections of Polyhexes." In *Renaissance Banff: Mathematics, Music, Art, Culture Conference Proceedings 2005*, edited by Reza Sarhangi and Robert V. Moody, pp. 427–434. Winfield, KS: Central Plain Book Manufacturing, 2005.

[7] Branko Grünbaum and G.C. Shephard, Tilings and Patterns, W.H. Freeman, New York, 1987.

[8] Peter Raedschelders. "Tilings and Other Unusual Escher-Related Prints." In *M.C. Escher's Legacy*, edited by Doris Schattschneider and Michele Emmer, pp. 230–243. Berlin: Springer-Verlag, 2003.

Creating the NAVIGATI Puzzle

Adrian Fisher

NAVIGATI was launched worldwide on January 31, 2006, in the British national newspaper *The Daily Mail*, and it gained an immediate and enthusiastic daily following. It first appeared on April 8, 2006, in a second British national newspaper, *The Guardian*. This article explains the design approach behind the puzzle concept.

Creating Puzzles for Different Markets

Different markets for two-dimensional puzzles require a different design approach to maximize their effectiveness, even though the change of emphasis may only vary slightly.

In a newspaper with its lifespan typically measured in minutes (and at most, 24 hours), the puzzle creator actively encourages you to write on the paper, as part of your involvement and pleasure, perhaps to pass the time on a commuter train, or during a coffee break at any moment of the day.

Alternatively, in a large-format hardcover book on fine paper with color photographs, the author appreciates that you would prefer to use the blunt end of a matchstick to trace your progress, so as not to spoil the book for future enjoyment by yourself or others.

Yet again, in the outdoor courtyard of a corn maze, the puzzle may be a Six Minute Maze, presented as a flat design eight feet square to walk upon. You use your own body to move about the design, with your position denoting your progress; every time you make a turn, you look down and experience further confusion since the puzzle no longer appears the same way up.

These are all different ways of presenting non-mechanical statically presented two-dimensional puzzles; when the puzzle has physical substance, typically in the form of plastic, durable printed cardboard or both, further different design considerations apply.

Creating Puzzles for Newspapers

Creating a new puzzle concept for newspapers and magazines is a distinct design challenge.

The most successful example of the genre is the Crossword Puzzle. The reader has a series of tasks that he can attempt in any order; there is a gratifying Virtuous Reward as he progresses, and previous successes provide extra clues for the final words that he has found most difficult to solve. The puzzle uses both sides of the brain, since on one side there is something strategic about gradually covering the space, and on the other side there is the verbal teasing of the clues, the stretching of the vocabulary, and the reinforcement of one's accurate spelling ability. Physically, each small achievement involves putting pen to paper, and making one's mark; at any stage, one can proudly see the state of progress at a glance.

Newsprint space is at a tremendous premium. To an editor, a crossword puzzle offers an impressive combination of a compact number of square inches occupied (including clues and yesterday's solution), a large number of minutes spent by readers per square inch (probably the highest of any square inch in the newspaper), and a method of developing strong readership loyalty and thus regular repeat sales.

Su Doku provides many of the same attributes as crossword puzzles, though without the verbal aspects that unleash so many pleasurable visual images and ideas while solving a crossword puzzle.

Yet it is the phenomenon of Su Doku's rapid worldwide success, especially since 2005, which has reminded newspapers of the circulation-boosting value of puzzles, and one can notice the extra space being devoted to puzzles of all kinds, in newspapers around the world.

Creating NAVIGATI

My design objectives when conceiving NAVIGATI were as follows.

1. It must have a compact size, maximum of 85 × 150 mm (3 × 5 inches).

2. It must have a distinctive name. NAVIGATI implies a navigation puzzle, something rather exotic from abroad that suggests something Italian, and ends in a vowel like all the latest and most popular puzzles.

3. It must have a distinctive logo—crucial for drawing the eye across the page—the signature indicates that it was personally created by a named designer.

4. It must have a distinctive visual format—like an ancient mariner's sea chart, hence the navigation circles and compass-like arrow indicators, choice of fonts, and the distinctive map-like black and white edging. It really stands out from other puzzles on a page.

5. I wanted a maze-like challenge. The puzzle involves decision-points, with its paths being imaginary hops through the air.

6. It cannot be immediately solved by eye, just by glancing at it (unlike plans of traditional hedge mazes, which the eye can solve by racing around the design in moments).

7. I wanted to maximize the puzzlement. The combination of distance and direction at every jump adds to the puzzlement and concentration required.

8. One has to write one's progress on paper, to keep track of where you have been. With the true route jumping all over the place, this is essential.

9. It must have a verbal dimension—hence extracting letters from your solution to make an anagram, and then solving the anagram.

10. Each solution word is a travel destination of some kind—such as a town, city, region, country, mountain, river, lake, or ocean.

11. It must have milestones. These provide reassurance and confirmation at various stages of your solution. They also allow you to tackle the puzzle in a different order if you wish, e.g., the last few moves first.

12. The clue letters around the edge form two words—not because they have to, but just because it's more fun that way.

13. It must have no black holes. (In astronomy, a *black hole* is somewhere you can get into, but can never escape.). By having no black holes within the puzzle, players never get stuck, and they can keep moving around the design indefinitely or until they solve the puzzle. Secondly, their challenge is to find the shortest solution.

14. It must have some white holes. (*White holes* are cells you cannot reach from the start, but that you will get caught up in, if you try and work out the puzzle backwards). White holes are the opposite of black holes, and are deliberately included in each puzzle. It's another aspect of the fun of the puzzle, and afficionado players may like to work out which of the cells are white holes.

15. It must have a format capable of providing thousands of different NAVIGATI puzzles—hence the combination of distance and direction. This achieves an immense amount of variety, and one can introduce or eliminate "short circuits" simply by adding or removing an arrow at the design stage.

16. It is crucial to have a solution with no black holes, of the predicted length, and be able to guarantee that there is only one solution. Otherwise there might be more than one valid set of anagram letters.[1]

17. I must be able to deliver a new puzzle consistently and in time, day after day. Otherwise, it cannot appear every single day in the national press!

---

[1]As much time was spent creating a rigorous computer testing procedure regardless of the size of the puzzle, as all other aspects of its puzzle content and graphic design.

18. It must be targeted to a particular newspaper and its readership.[2]

19. The choice of planet Earth is a suitable source of a sufficient number of well-recognized six-letter destination words. We're not sure that this choice of planet will stand up for more than a few hundred six-letter destinations! But so far, so good. And it was the best available planet at the time.

The NAVIGATI Puzzle

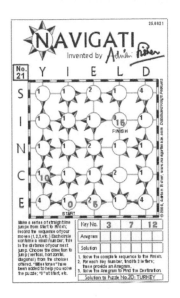

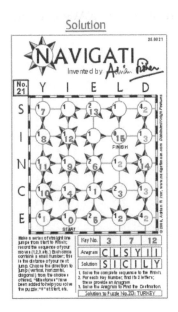

Figure 1. A sample of the NAVIGATI puzzle.

The result is the NAVIGATI puzzle, which combines solving a maze network with gathering letters to generate an anagram. The published rules are as follows:

---

[2] The initial 5 × 5 format and the six-letter anagram were selected for the particular readership of a first British newspaper. A second British newspaper chose a 7 × 7 format with an eight-letter anagram. The puzzle can be easily devised to specific market requirements.

Make a series of straight line jumps from "Start" to "Finish," and record the sequence of your moves (1, 2, 3, etc.). Each circle contains a small number; this is the distance of your next jump. Choose the direction to jump (vertical, horizontal, diagonal), from the choices offered. "Milestones" have been added to help you solve the puzzle—"0" at "Start," etc. Thus:

1. solve the complete sequence to the "Finish";

2. for each "Key Number," find its two letters—these provide an anagram;

3. solve the anagram to find the destination.

A sample of the NAVIGATI puzzle is shown in Figure 1.

Crazy Elephant Dance

Markus Götz

Do you know the wonderful puzzle "Elephant Spin Out" by Binary Arts? It is a mechanical sliding puzzle, in which you have to slide a red bar out of a green shell. On top of the red bar there are attached elephants that can be rotated. The mechanism is designed so that the elephants all have to point to the exit before the red bar can be slid out completely. Figure 1 shows a picture of the puzzle.

Figure 1. "Elephant Spin Out" by Binary Arts. (See Color Plate IX.)

165

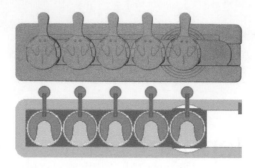

Figure 2. A mechanical version of "Crazy Elephant Dance" using five elephants (top), and the most interesting interlocking layers of the puzzle
(bottom).

Each elephant has exactly two orientations, which define a binary system. According to its solution algorithm it is equivalent
to the "Chinese Rings" puzzle or "The Brain." My idea was to extend this puzzle using three orientations for each elephant instead
of two, with the intention of lifting the puzzle "Elephant Spin Out"
from a binary version to a ternary version. Figure 2 shows the
physical implementation of this ternary version. Here all elephants
are in their starting positions (pointing upwards).

The mechanism of the puzzle allows the following rules:

1. each elephant has three orientations: pointing upwards, pointing to the right, and pointing downwards;

2. an elephant can be rotated only at the position of the circular
   cutoff;

3. an elephant can be rotated from pointing upwards to pointing
   to the right (or vice versa) only when the elephant to its right
   is pointing downwards;

4. an elephant can be rotated from pointing downwards to pointing to the right (or vice versa) only when the elephant to its
   right is pointing upwards;

5. an elephant can exit (or enter) the frame only when pointing
   downwards.

Since the objective is to slide all the elephants out of the frame, you
have to turn them so they all point downwards. The major differ-

ence between "Crazy Elephant Dance" and the other puzzles mentioned above is that in this one there is not just one path from the starting point to the finish, but there are crossings, wrong paths, and dead ends.

Calculating the Number of Rotations

From a mathematical point of view, it is interesting to know the number of steps required to solve the puzzle (by the way, it is a nice example for students to use induction). Let's assume we have given the puzzle "Crazy Elephant Dance" with $n \geq 1$ elephants. The elephants are numbered in ascending order from 1 to n, counting from right to left. We define $S(n)$ to be the number of rotations required to go from the starting position (all elephants pointing upwards) to the finishing position (all elephants pointing downwards), so that all the elephants can be slid out of the frame. Sliding all the elephants in order to match the next elephant to its rotating position is not counted in the number of moves.[1]

First we will develop a recursive formula for $S(n)$, and then by guessing an explicit formula for $S(n)$, we can use induction to prove that it is correct.

To get a recursive formula for $S(n)$ we introduce another number: let $R(n)$ be the number of rotations needed to change the orientation for elephant number n from up to down (or vice versa), while the elephants with numbers 1 to $n-1$ are all pointing down. Figure 3 shows both definitions $S(n)$ and $R(n)$.

For $n = 1$ we can state the numbers: $S(1) = 1$ and $R(1) = 1$. For $n \geq 2$ we can combine S and R to get a formula for $S(n)$ (see Figure 4 for the individual steps):

$$S(n) = S(n-1) + R(n), \quad n \geq 2. \tag{1}$$

The most interesting thing about this is that we have been able to reduce the index of S from n to $n-1$. But by doing so we have to use the second unknown R. This means that we now need a formula for that as well. With the detailed steps shown in Figure 5

---

[1]In fact, if you regard sliding the elephants as one move, then you have to double the calculated number $S(n)$, because starting with the first elephant you have to do one slide after each rotation either to align the next elephant to the rotating position or to slide all of them out of the tray.

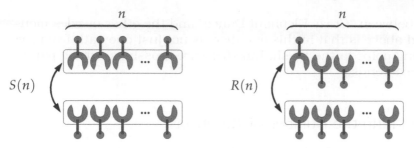

Figure 3. Definition of the number of moves $S(n)$ and $R(n)$.

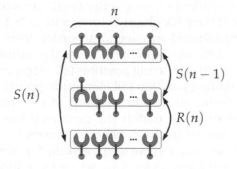

Figure 4. Individual steps for getting formula (1).

we can write:

$$
\begin{aligned}
R(n) &= 1 + R(n-1) + 1 + R(n-1) \\
&= 2R(n-1) + 2, \quad \text{for } n \geq 2.
\end{aligned}
\tag{2}
$$

Using $R(1) = 1$ it is easy to calculate $R(2) = 2R(1) + 2 = 4$, and with that $S(2)$, which we will use later:

$$
S(2) = S(1) + R(2) = 1 + 4 = 5.
$$

Formula (2) uses only R. This means that we could now try to find an explicit formula for R and use it in (1) to get an explicit one for S. But this would require guessing an explicit formula and using induction twice (once for R and once for S). To minimize the work, we will focus on getting an explicit formula for S.

To get rid of the auxiliary variable R, we use (1) twice for $R(n)$ and $R(n-1)$ in (2) with $n \geq 3$. For the time being this leads to a recursive formula for S only:

$$
S(n) = 3S(n-1) - 2S(n-2) + 2, \quad \text{for } n \geq 3.
\tag{3}
$$

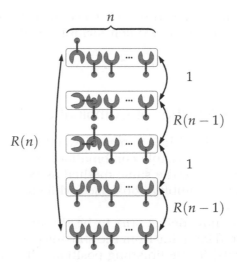

Figure 5. Individual steps for getting formula (2).

To get an explicit formula for S out of (3) you either have to read the first chapter of [2] and use the techniques presented there to derive it, or you have to guess it (or else it is told to you) and prove it by using induction. We will use the second approach, although the first one is used to verify the following result.

The explicit formula for $S(n)$ is

$$S(n) = 3 \cdot 2^n - 2n - 3, \quad \text{for } n \geq 1. \tag{4}$$

We will prove it by using induction:

Proof: $n = 1$: For $n = 1$ the formula (4) is valid: $1 = 3 \cdot 2^1 - 2 \cdot 1 - 3$.
$n = 2$: For $n = 2$ the formula (4) is valid: $5 = 3 \cdot 2^2 - 2 \cdot 2 - 3$.
$n \geq 3$: Assuming that the formula $S(k) = 3 \cdot 2^k - 2k - 3$ already holds for all $k = 1, \ldots, n-1$, we can then write:

$$
\begin{aligned}
S(n) &\overset{(3)}{=} 3S(n-1) - 2S(n-2) + 2 \\
&= 3\left(3 \cdot 2^{n-1} - 2(n-1) - 3\right) - 2\left(3 \cdot 2^{n-2} - 2(n-2) - 3\right) + 2 \\
&= 9 \cdot 2^{n-1} - 6n + 6 - 9 - 6 \cdot 2^{n-2} + 4n - 8 + 6 + 2 \\
&= (9 \cdot 2^{n-1} - 3 \cdot 2^{n-1}) - 2n - 3 \\
&= 3 \cdot 2^n - 2n - 3.
\end{aligned}
$$

This proves that also in the case $n \geq 3$ formula (4) holds.   □

By the way, now we can use (4) twice in (1) for the expressions $S(n)$ and $S(n-1)$ to get an explicit formula for R:

$$R(n) = 3 \cdot 2^{n-1} - 2.$$

Calculating the Number of Combinations Used for Solving

Because every elephant has a total of exactly three possible orientations, there are 3^n possible combinations for the orientations of the n elephants. Using the same method as before, we can calculate the number of combinations used while solving the puzzle in the shortest way.

To do so, we now define $s(n)$ to be the number of *90-degree-rotations* required for going from the starting position (all elephants pointing upwards) to the finishing position (all elephants pointing downwards). Counting only the $90°$-rotations for R leads us to the definition of $r(n)$. As a result we can state the initial numbers $s(1) = 2, r(1) = 2$. Convincing ourselves that the formulas (1), (2), and (3) also hold in this case, we can calculate $r(2) = 2 \cdot r(1) + 2 = 6$, enabling us to get $s(2) = s(1) + r(2) = 8$. Having these initial values for $s(n)$, $n = 1, 2$, we can prove the following formula using induction:

$$s(n) = 2^{n+2} - 2n - 4, \quad n \geq 1.$$

Each of the $s(n)$ $90°$-rotations leads us to a new combination. This means that in order to get all the combinations using n elephants, we only have to add the first one (= starting position):

$$s(n) + 1 = 2^{n+2} - 2n - 3.$$

It is interesting to see that the base of the leading summand is only 2 and not 3. The number of combinations that are not used for the solution but exist nonetheless, leading the puzzler to wrong ways and dead ends, is the difference between $s(n) + 1$ and 3^n:

$$w(n) := 3^n - 2^{n+2} + 2n + 3.$$

Some Numbers

To get a feeling for the numbers $S(n)$ with increasing n, see Table 1.

n	elephants	1	2	3	4	5	6	7	8	9	10
$S(n)$	rotations	1	5	15	37	83	177	367	749	1515	3049
$s(n)+1$	combinations	3	9	23	53	115	241	495	1005	2027	4073
$w(n)$	wrong combinations	0	0	4	28	128	488	1692	5556	17656	54976

Table 1. Rotations and combinations for $n = 1, \ldots, 10$.

As an example, if you want to solve the "Crazy Elephant Dance" with $n = 7$ elephants and you succeed in doing one rotation within one second (and the necessary shift of the slide with the elephants so as to align the next elephant in the rotating position), then it will take you six minutes and seven seconds. And then it will take you the same time again for resetting the puzzle to the starting position.

Give It a Try

If you would like to try this puzzle you are invited to do so. Use the website [1] where you can start the Java applet by means of an indicated button. In the applet you can play with the puzzle using $n = 1, \ldots, 7$ elephants. I implemented a nice feature that will do one correct next move for you leading to the goal. For demonstration purposes you can also run the solving routine that works move by move until the puzzle is solved.

Mechanism

When I started to design the puzzle and to find a working mechanical version that implements the rules 1–5 stated above, my first sketches of the mechanism were very complicated. But after some improvements, it became more and more simple and elegant. Nevertheless, it is definitely more complex than the original "Elephant Spin Out." The puzzle was designed to be lasercut, the individual layers being glued together afterwards. Figure 6 shows all seven layers of the puzzle. Participating in the Nob Yoshigahara Puzzle Design Competition in 2005 at the IPP25 in Helsinki, this design was given an "Honorable Mention" by the jury. Unfortunately, no company has so far decided to produce this puzzle for market and so it is not available commercially.

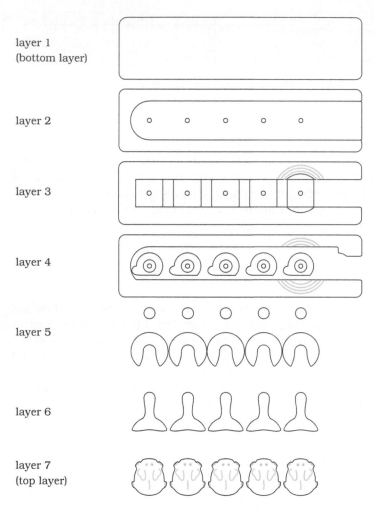

layer 1
(bottom layer)

layer 2

layer 3

layer 4

layer 5

layer 6

layer 7
(top layer)

Figure 6. All seven layers of the puzzle: layer 1 is the bottom plate; layer 2 is the slide with the five axes of the elephants; layer 3 has the restriction that allows the elephants to be rotated only at the second position from the right (rule 2) and allows them to have four orientations at the other places (partly fulfills rule 1); layer 4 prevents elephants from pointing to the left (now rule 1 is fulfilled) and allows elephants to exit the tray only while pointing downwards (rule 5); layer 5 is the interlocking layer that realizes rules 3 and 4; layer 6 is the holder for the circles in layer 5, which also act as the trunks of the elephants; layer 7 is the top layer in the shape of an elephant, which imparts the theme to the whole puzzle.

Bibliography

[1] Markus Götz. "Crazy Elephant Dance." Available at http://
www.markus-goetz.de/puzzle/0019.html, 2008.

[2] Herbert S. Wilf. *generatingfunctionology*, Third Edition. Wellesley, MA: A K Peters, 2005. (Second Edition available at http://
www.math.upenn.edu/~wilf/DownldGF.html.)

The Two Ovals-to-Table Story

Serhiy Grabarchuk

This well-known classical puzzle, which is based on the transformation of two oval stools into a circular table top, has a long and interesting history. More than 180 years ago, in 1821, John Jackson posed in his book, *Rational Amusement for Winter Evenings* [2], the question of how to transform a circle into two congruent hollow ovals, and he proposed an eight-piece solution. Jackson's Table and Oval Stools puzzle is depicted in Figure 1. Can you discover how to reach the eight-piece solution?

Then at the beginning of the twentieth century, this puzzle attracted the attention of Sam Loyd, who was America's greatest puz-

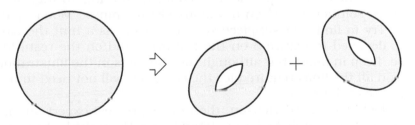

Figure 1. Jackson's Table and Oval Stools puzzle.

175

PROPOSITION—Divide the two ovals into the fewest possible number of pieces which will fit together and form one large circular piece.

Figure 2. Loyd's "An Old Saw with New Teeth" puzzle as it appeared in his *Cyclopedia of 5000 Puzzles, Tricks & Conundrums.*

zle creator. In Loyd's legendary *Cyclopedia of 5000 Puzzles, Tricks & Conundrums* [3], the puzzle was posed—under the name "An Old Saw with New Teeth"—as the problem of dissecting two congruent ovals, each with an oval-shaped "hand hole," into the smallest number of parts that can be combined to form a circular table top. The puzzle is shown in Figure 2.

Loyd published the puzzle as a contest, showing an improved six-piece solution based on a famous Great Chinese Monad pattern. Try to find that six-piece solution, using as a hint the patterns depicted in Figure 3 on the two ovals and on the resulting circle. Keep in mind that although the diagrams in the illustration contain all the lines required for the cuts, you will not need to use every one of these lines.

In 1997 Greg Frederickson, the world's greatest expert on dissection puzzles, published an outstanding book on the history and achievements of this old and comprehensive field of recreational

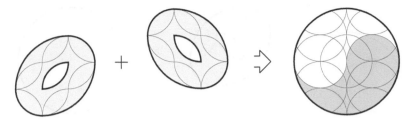

Figure 3. Pattern for six-piece solution based on the Great Chinese Monad.

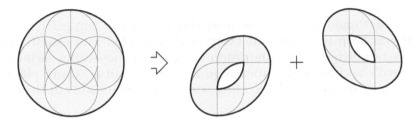

Figure 4. Diagram for five-piece solution.

math—*Dissections: Plane & Fancy* [1]. Chapter 15 of his book is devoted exclusively to dissections with curved figures. He describes both the Loyd puzzle and some variations of it with two hollow ovals. In his book, Frederickson shows a different and quite novel six-piece solution to Jackson's ovals.

Since the first publication of Jackson's puzzle, there have been numerous attempts to improve on the six-piece solution. Finally, in March of 2004, the author discovered several new six-piece solutions, two basic solutions containing just five (!) pieces each, more than a dozen different modifications of these basic five-piece solutions, and proofs that in a mathematical sense there is an infinite number of five-piece solutions. In every one of them, one piece is flipped over. Finding any of the five-piece solutions is not an easy task. Try to discover one of the simplest examples of them. Can you do this? (*Hint*: The diagrams in Figure 4 contain all the lines required for your cuts. Of course, not all of the lines are needed. And keep in mind that in a five-piece solution, one piece can be flipped over.)

Last but not least, there is a variation of the hollow-ovals-to-circle puzzle posed by Sam Loyd in his attempts to find a solution with the smallest number of pieces. It also is based on the Great

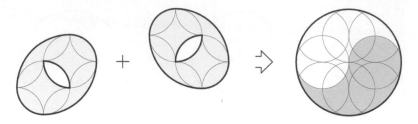

Figure 5. Loyd's variation with four pieces.

Chinese Monad pattern described above, and you can easily solve
it if you keep in mind that every oval is divided into exactly the
same two pieces. The illustration in Figure 5 shows this variation
of Loyd's and all of the lines required for the cuts. Again, please
remember that you will need just some of these lines to cut the two
ovals.

Solutions

The solutions to the puzzles with eight, six, and five pieces, and the
four-piece solution to another variation by Sam Loyd, are shown
in Figures 6, 7, 8, and 9.

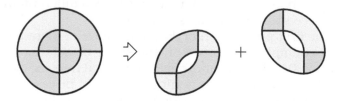

Figure 6. Eight-piece solution by John Jackson, 1821.

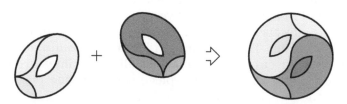

Figure 7. Six-piece solution by Sam Loyd, 1901.

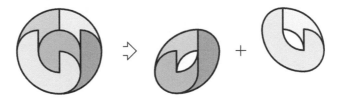

Figure 8. Five-piece solution by the author, 2004.

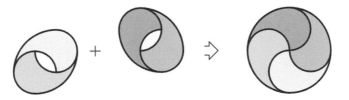

Figure 9. Another variation by Sam Loyd with four-piece solution.

Acknowledgment

This article was specially prepared for the 25th International Puzzle Party, Helsinki, Finland, July 22–25, 2005.

Bibliography

[1] Greg Frederickson. *Dissections: Plane & Fancy*. Cambridge, UK: Cambridge University Press, 1997.

[2] John Jackson. *Rational Amusement for Winter Evenings*. London: Barry and Son, High-Street, 1821.

[3] Sam Loyd. *Cyclopedia of 5000 Puzzles, Tricks & Conundrums*. New York: Lamb Publishing Company, 1914.

Folding Regular Heptagons

Thomas C. Hull

The number seven has been historically problematic in the study of constructing regular polygons. Using a straightedge and compass to draw a regular heptagon with mathematical precision is, alongside angle trisections and cube doublings, one of the classic Greek geometric construction problems. Mathematicians have known for a few hundred years now that such constructions are impossible to do with these tools. On the other hand, for less than a hundred years mathematicians have known that origami (paper folding) was capable of conquering these construction conundrums.

As tribute to the seventh Gathering for Gardner, we will present the theory behind origami regular heptagon construction as well as how to implement it in two different ways.

The Theory Behind the Folds

As far as anyone knows, the first person to develop a folding method for the regular heptagon and commit it to paper was Benedetto Scimemi [13]. His approach uses a standard analysis of the heptagon in the complex plane, as can be seen in Andrew Gleason's paper [6].

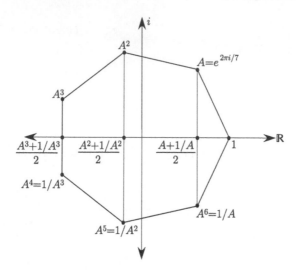

Figure 1. The seventh roots of unity in the complex plane form a heptagon.

The idea is to think of each of the regular heptagon's corners as points in the complex plane corresponding to the seventh roots of unity, that is, solutions to the complex equation $z^7 - 1 = 0$. Figure 1 shows the seventh roots of unity in the complex plane forming a heptagon. Of course, $z = 1$ will be one such root, but so will $e^{2\pi i/7}$, since raising this number to the seventh power is the same as adding the angle $2\pi/7$ to itself seven times, giving us $e^{2\pi i} = 1$. If we let $A = e^{2\pi i/7}$, we see that any power of A will also be a solution of $z^7 - 1 = 0$. Thus, we also have solutions

$$A^2 = e^{2\frac{2\pi}{7}}, \ A^3 = e^{3\frac{2\pi}{7}}, \ A^4 = e^{4\frac{2\pi}{7}}, \ A^5 = e^{5\frac{2\pi}{7}}, \ A^6 = e^{6\frac{2\pi}{7}},$$

and $A^7 = 1$ gets us back to where we started and no more roots will be found. These seven points constitute seven equally-spaced points on the unit circle in the complex plane.

Our job, then, is to locate these points in a sheet of paper via folding. From a folding point of view, if we could just locate the point A, then assuming we already have the point 1 constructed on the real axis, we would have the angle $2\pi/7$ constructed, which could then be copied around the origin to produce the rest of the heptagon. From a mathematical point of view, we are really trying to solve the equation $z^7 - 1 = 0$, and factoring out the root given by

$(z-1)$ gives us

$$\frac{z^7-1}{z-1} = z^6 + z^5 + z^4 + z^3 + z^2 + z + 1 = 0. \tag{1}$$

We can convert this 6th degree equation to a more simple one using the symmetry of our heptagon. Since $A = e^{2\pi i/7} = \cos(2\pi/7) + i\sin(2\pi/7)$,

$$\frac{1}{A} = \frac{\overline{A}}{A\overline{A}} = \frac{\overline{A}}{\cos^2(2\pi/7) + \sin^2(2\pi/7)} = \overline{A} = A^6.$$

Therefore, $A + 1/A = A + \overline{A} = 2\cos(2\pi/7)$. This can be seen geometrically as well; $(A + 1/A)/2$ is the midpoint of the line segment connecting A and $1/A$, which is on the real axis at $\cos(2\pi/7)$. (This is shown in Figure 1.) Since $\cos(2\pi/7)$ will give us the angle we want, this is really all we need, so we'll turn our focus to the quantity $A + 1/A$.

We can also show that $A^5 = 1/A^2$ and $A^4 = 1/A^3$ (again, see Figure 1). This means that equation (1), if we let $z = A$, becomes

$$\frac{1}{A} + \frac{1}{A^2} + \frac{1}{A^3} + A^3 + A^2 + A + 1 = 0. \tag{2}$$

By expanding out $(A + 1/A)^2$ and $(A + 1/A)^3$ and simplifying, we get that

$$A^2 + \frac{1}{A^2} = (A + 1/A)^2 - 2 \text{ and } A^3 + \frac{1}{A^3} = (A + 1/A)^3 - 3(A + 1/A).$$

Substituting these into equation (2), we get

$$(A + 1/A)^3 + (A + 1/A)^2 - 2(A + 1/A) - 1 = 0.$$

Therefore, $A + 1/A = 2\cos(2\pi/7)$ is a solution to equation

$$z^3 + z^2 - 2z - 1 = 0. \tag{3}$$

Similar machinations show that that the other two roots of equation (3) are $2\cos(4\pi/7)$ and $2\cos(6\pi/7)$. In other words, equation (3) is the one we want to solve via origami.

Folding It

That origami can solve general third-degree equations was first discovered by the Italian mathematician Margherita Beloch in the 1930s [3]. She did this by utilizing an origami move that had not been previously considered: fold two points to two different lines simultaneously. Beloch demonstrated this by outlining a folding method for constructing $\sqrt[3]{2}$, which is needed to solve the classic Greek problem of doubling the cube. This "two points to two lines" origami move can also by utilized to trisect arbitrary angles (see [8, 10]). It will also allow us to construct a regular heptagon.

To see what fold will solve equation (3), and thus construct $2\cos(2\pi/7)$, imagine our piece of paper is the infinite real plane. (We will see how to do this in an actual square of paper in a moment.) Label the points $P_1 = (0,1)$ and $P_2 = (-1,-1/2)$, and let L_1 be the x-axis and L_2 be the y-axis. We then perform Beloch's origami move: fold the plane so that P_1 lands on L_1 while at the same time making P_2 land on L_2. Let t be the place on the x-axis where P_1 lands; call this point $P_1' = (0,t)$. (See Figure 2.)

Let's figure out what value t must be. We can do this by determining the equation of our crease line (the dashed line in Figure 2). The segment $P_1 P_1'$ has slope $(1-0)/(0-t) = -1/t$, and our crease line will be the perpendicular bisector to this segment. Thus, our

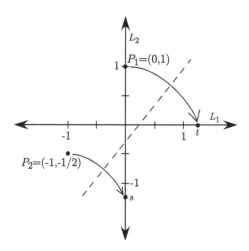

Figure 2. Folding two points to two lines to construct $t = 2\cos(2\pi/7)$.

crease has slope t and the crease must pass through the midpoint of P_1P_1', which is $(t/2, 1/2)$. Using the point-slope formula for a line, we have that an equation for the crease line is

$$y - 1/2 = t(x - t/2) \implies y = tx - t^2/2 + 1/2.$$

On the other hand, if we let $P_2' = (0, s)$ be the point where P_2 lands on the y-axis, then the segment P_2P_2' has slope $(2s + 1)/2$ and midpoint $(-1/2, (2s - 1)/4)$. Thus, another formula for our crease line is

$$y = \frac{-2}{2s + 1}x - \frac{1}{2s + 1} + \frac{2s - 1}{4}.$$

Since these two lines are the same, they must have the same slope. This gives us $s = -(t + 2)/(2t)$. But the constant terms of these two line equations must also be equal, that is,

$$-t^2/2 + 1/2 = -1/(2s + 1) + (2s - 1)/4.$$

If we substitute $s = -(t + 2)/(2t)$ into this and simplify, we'll get a single equation in t:

$$-\frac{t^2}{2} + \frac{1}{2} = \frac{t}{2} - \frac{t + 1}{2t} \implies t^3 + t^2 - t = t + 1 \implies t^3 + t^2 - 2t - 1 = 0.$$

Lo and behold, we see that t is a solution to equation (3)! Since the other two roots of equation (3) are negative, we have that t must equal $2\cos(2\pi/7)$.

We really wanted to construct the angle $2\pi/7$, but now this will be easy. Since we have constructed $2\cos(2\pi/7)$ on the x-axis, all we would need to do is fold a line at the point t perpendicular to the x-axis and use this to make a right triangle whose base is $2\cos(2\pi/7)$, whose side is the line perpendicular at t, and whose hypotenuse (beginning at the origin) is of length 2. The angle this triangle makes at the origin would then be $2\pi/7$.

All that's left is to implement the origami move in Figure 2 on an actual sheet of paper and do the proper manipulations described in the previous paragraph to create our $2\pi/7$ angle. There are many ways to achieve the coordinate system in Figure 2 on a sheet of paper. We present one developed by Robert Geretschläger in [5]. Take a square piece of paper and assume that the origin is in the center and that the side of the square is four units long. The x- and y- axes can then be constructed by folding the paper in half

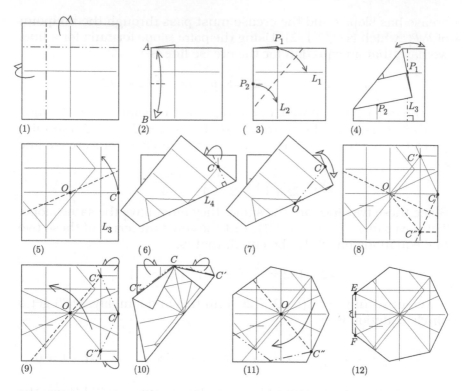

Figure 3. Folding a regular heptagon from a square piece of paper.

in both directions, and the points $P_1 = (0,1)$ and $P_2 = (-1,-1/2)$ can be found as shown in steps (1)–(2) in Figure 3. The "Beloch move" of folding two points to two lines can then be performed as seen in step (3). Since at this stage, the points P_1 and P_2 are on the edges of the folded paper, this move is relatively easy to perform, but it does require bending the paper over and carefully lining up the points before pressing the crease flat.

The other steps in the folding sequence in Figure 3 (which is just a simplification of Geretshläger's in [5]) are explained as follows:

1. Step (4). After folding P_1 onto the x-axis (line L_1), its location will be at $t = 2\cos(2\pi/7)$. Folding the vertical line L_3 that passes through this point is making the side of the right triangle described previously.

2. Step (5). Then unfold everything. We want to make the hypotenuse of length 2 for the right triangle. Since the segment marked OC is of length 2, we can simply fold the paper so that C is placed on line L_3 and the crease passes through O. This is equivalent to using a compass to draw an arc of a circle centered at O and with radius OC. Where C lands on L_3 (call it C') will be the top point of our right triangle. In other words, $\angle C'OC = 2\pi/7$.

3. Step (6)–(7). These simply mark where the point C' is and crease the line OC'.

4. Step (8). Then unfold everything and repeat steps (5)–(7) on the bottom half of the paper to produce the point C''. So far we have constructed the points equivalent to A and A^6 in Figure 1.

5. Step (9)–(12). The rest is simply copying the angles $\angle C'OC$ and $\angle C''OC$ around the origin, while at the same time creasing the other sides of the heptagon.

Folding a Modular Heptagon

Folding a regular heptagon from a single sheet of paper is quite challenging. The folds needed are not trivial, and slight inaccuracies can result in a rather irregular-looking heptagon. With practice, however, it can be done very precisely.

However, we could also use what we learned to make a modular origami heptagon, and one could argue that this would result in better accuracy. *Modular origami* is where we fold many pieces of paper to make "units" that are then linked together to form a larger, usually geometric object.

There are many ways to modify the folding method for a regular polygon to a modular unit for the same kind of polygon. All that's needed is the construction of the proper angle. We chose to create a heptagon version of Robert Neale's classic Magic Ring (aka Pinwheel-Ring) from [12]. Neale, a well-known magic author whose work has appeared in a number of Martin Gardner columns, is also a famous origamist. His Magic Ring is especially popular because it can transform from an eight-sided pinwheel shape to an octagon

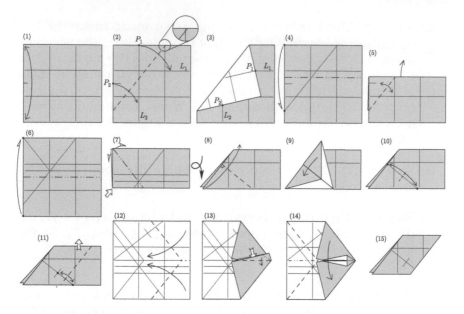

Figure 4. Folding the unit for a modular heptagon ring.

ring. His sliding and locking mechanisms are especially simple and a natural choice for adaptation to other polygons.

Figure 4 shows the folding sequence for the heptagon ring unit. This is, of course, much more complicated then Neale's original. The construction of the $2\pi/7$ angle is done using a different coordinate system than the single-sheet version of Figure 3. For the modular version we choose to let the paper be three units to a side where the upper right-hand corner has coordinates $(2,1)$ and the lower left-hand corner is $(-1,-2)$. Thus, the first thing we need to do is to crease the square paper into *thirds* horizontally and vertically. There are a number of ways to do this with mathematical precision, but origamists usually find it easier to just roll the paper into an S-shape and flatten carefully into thirds.

One can then see that step (2) in Figure 4 is the same fold as before: folding $P_1 = (0,1)$ onto the L_1 (the x-axis) while at the same time bringing $P_2 = (-1,-1/2)$ onto L_2 (the y-axis). This creates the desired angle of $2\pi/7$. The rest of the folding sequence involves moving this angle into the proper place so as to create the structure needed for Neale's unit.

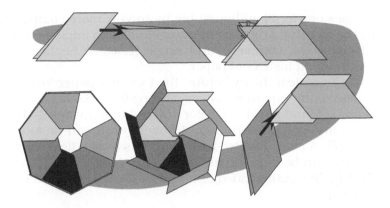

Figure 5. Locking the heptagon ring units together.

Seven units need to be folded to form the heptagon ring, and at 15 steps that require a good amount of precision folding, this can be a very time-consuming task. But this can be made much easier as follows: Take care to create one very precise unit using the steps in Figure 4. Then use this as a *template* to make your other units. After all, the finished unit is really just a square piece of paper folded in half and then turned into a parallelogram with two parallel folds at an angle of $2\pi/7$ from the bottom. These two parallel folds can be made very accurately by using the first unit as a guide. One can hold the first unit on top of the in-progress unit to make the needed parallel folds, or partially unfold the first unit and find more clever ways to transfer the angles onto the new unit. Doing this also has the advantage of making units that do not have all the auxiliary creases marring the finished surface.

Instructions for putting the units together to form the ring-pinwheel can be seen in Figure 5. Each unit simply "hugs" a neighboring unit, and you want to make them hug each other tightly. Putting the last ones in correctly is something of a puzzle, as they do want to slide along their neighbors and the layers of paper become tricky.

Conclusion

Scimemi's [13] and Geretschläger's [5] heptagons have been the standard ones known in the origami community. Roger Alperin also provides one in [2], although his method is more complicated. He uses the fact that origami can trisect angles to implement Gleason's straightedge, compass, and angle trisector heptagon construction in [6] on a sheet of paper.

The mathematical theory behind the fact that paper folding can solve general cubic equations is very interesting. Basic expositions of this can be found in [11, Chapter 10] and in Activities 4–6 of [9]. A fun nuts-and-bolts description of how to do it can be found on the web at [7]. A more detailed look at all this from a field extension point of view can be found in another paper of Alperin's [1] as well as in the highly recommended text *Galois Theory* by David Cox [4, Section 10.3].

Bibliography

[1] R. C. Alperin. "A Mathematical Theory of Origami Constructions and Numbers." *New York Journal of Mathematics* 6 (2000), 119–133.

[2] R. C. Alperin. "Mathematical Origami: Another View of Alhazen's Optical Problem." In *Origami³: Third International Meeting of Origami Science, Mathematics, and Education*, edited by T. Hull, pp. 83–93. Natick, MA: A K Peters, 2002.

[3] M. P. Beloch. "Sul metodo del ripiegamento della carta per la risoluzione dei problemi geometrici." *Periodico di Mathematiche* Series 4, Vol. 16 (1936), 104–108.

[4] D. Cox. *Galois Theory*. Hoboken, NJ: John Wiley & Sons, 2004.

[5] R. Geretschläger. "Folding the Regular Heptagon." *Crux Mathematicorum* 23:2 (1997), 81–88.

[6] A. M. Gleason. "Angle Trisection, the Heptagon, and the Triskaidecagon." *The American Mathematical Monthly* 95:3 (1988), 185–194.

[7] K. Hatori. "Origami Construction." Available at http://origami.ousaan.com/library/conste.html, 2003.

[8] T. Hull. "A Note on 'Impossible' Paperfolding." *The American Mathematical Monthly* 103:3 (1996), 242–243.

[9] T. Hull. *Project Origami: Activities for Mathematics Classes.* Wellesley, MA: A K Peters, 2006.

[10] J. Justin. Described in *British Origami* 107 (1984), 14–15.

[11] G. E. Martin. *Geometric Constructions.* New York: Springer, 1998.

[12] R. Neale and T. Hull. *Origami, Plain and Simple.* New York: St. Martin's Press, 1994.

[13] B. Scimemi. "Draw of a Regular Heptagon by the Folding." In *Proceedings of the First International Meeting of Origami Science and Technology,* edited by H. Huzita, pp. 71–77. Ferrara, Italy: PUBLISHER, 1989.

Combinatorial Philosophy

Kate Jones

Philosophy—a love of knowledge, learning, understanding, wisdom. A search for the truth about the world, the universe, reality, existence. A system for organizing the information and theories mankind has accumulated through the ages to explain existence and all it contains.

This definition of philosophy may not be the official, academic one, but it's the one I'll use for this outing.

In his intellectually intimate soul trek, *The Whys of a Philosophical Scrivener*, Martin Gardner leads us through a gamut of thought systems contrived by the most significant thinkers of the last few millennia. He identifies each system by its appropriate "*-ism*" and tells why he does not align himself with it. After examination of the whole palette, Martin "confesses" that he is a fideist, a Platonic mystic, a social democrat, and a believer in immortality. If anyone deserves immortality, in the way he defines it, Martin surely does.

About the perpetual tug-of-war between government planning and private enterprises—between altruism and egoism—in economic systems, Martin likes to say that "there may be no best way. There may be many ways, equally good. . . ."

Combinatorial puzzles—tiling sets of polyform shapes and color-differentiated polygons displaying all the permutations of a certain type and enabling assemblies in countless ways and with special conditions or criteria for a completed array. The types of "recreational mathematics" Martin Gardner popularized in his *Scientific American* columns, notably pentominoes and MacMahon's three-color squares.

I have described elsewhere ("Those Peripatetic Pentominoes," in *The Mathemagician and Pied Puzzler*, A K Peters, 1998) how Martin Gardner's writings on pentominoes directly influenced what has turned out to be the major part of my life's work: the development, production, and marketing of combinatorial puzzle sets. What remains to be said is how such puzzles can serve as paradigms for theories of "all there is."

We can observe all around us that matter, energy, life forms, and their aggregations tend toward organized groupings or systems, by some dynamic yet to be fully understood, although crystals and fractals give us hints. Systems grow from unity, or singularity, to greater and greater complexity. For example, polyforms, whether of squares, equilateral triangles or hexagons (the regular polygons that readily tile the plane), or more elaborate combinations, increase by leaps and bounds in the number of different shapes formed when the number of cells increases. From the 12 famous pentominoes (5 squares) to 35 hexominoes, 108 heptominoes, 369 octominoes—and that is just eight units! In how many more steps will the possible permutations exceed the number of atoms in the universe? Yet each shape is a distinct, unique entity unto itself.

The number of variations for each larger order builds on the previous set—expands hierarchically from the preceding level. So does the mind grow and organize its information, its nonstop flow of new sensory input. Each new bit of data gets integrated into the existing inventory, sometimes fastidiously, as in the scientific method, sometimes ramshackle, like a cluttered attic. All this experiential information serves as fodder for decision making on a moment-to-moment basis for survival, security, goal achievement, which goes beyond the mere autonomous instinctual apparatus. The higher mental functions, abstract thought, "thinking about thinking," allow the mind to form complex operational rules for itself and to exert self-correction. This higher level of brain function can override many of the built-in, animal-level behaviors. Let's call this ability "free will."

In many situations, when pressed for a decision, we may face several equally valid options. "There may be many ways, equally good ...." By some unfathomably complex set of algorithms, the mind will weigh the multitude of parameters and pop out an optimum answer, based on its unique context. Unpredictable? Probably. Especially from women, whose value assignments tend to be weighted in more complicated ways than men's. Shall we call this "intuition"? The workings of the subconscious? Causality? I would not call it determinism.

The 12 pentominoes can form a 6x10 rectangle in 2,339 distinct ways. If there is a God, and if God knows everything, past, present, and future, then God is aware of the infinitude of potentialities of how our world could develop, but would not necessarily know which convoluted strand or path, among all the infinite forks in the road, the world will take. We can know all 2,339 pentomino solutions, but not know which one of them the next solver will find.

Societies, or human groupings, accumulate their cultural and regulatory structures from the gradual combinatorial melding of their distinct individual units. Some pieces of this "puzzle" are bigger than others and have more influence, weight, or power. The blood-spattered history of the human race bears witness to an absence of organizing intelligence in accommodating all the pieces of the puzzle into a harmonious unity. America's founding fathers came the closest by setting forth a system to secure "life, liberty, and the pursuit of happiness." Shown in Figure 1 are some puzzles with all different "members" or "citizens" that have nevertheless been joined into a beautiful composition without having to demolish, mutilate, or throw away any of its members. I frequently use this example to defend individual rights against totalitarians: the point is that the mutual good of both polar opposites—the group versus the individual—can be achieved without loss or harm to either. One just needs to find for each member a suitable niche. There is no "redistribution of wealth" here. No one exploits any one else, not the rich soaking the poor, not the poor draining the rich. None needs to be diminished for the sake of any other.

Since there are "many ways, equally good" to reach a goal, we can sometimes discover, among the many ways, some that are outstanding for various reasons: beauty of symmetry, balance of coloration, robustness of cohesion, flexibility for change. Figure 2 shows an easy puzzle that readily assembles into other patterns

Figure 1. Puzzles with different members joined into compositions.

(with acknowledgment to Dr. Alan Schoen, the inventor of patented rhombic circle tilings).

And Figure 3 shows a very complex set of curvy shapes with hundreds of relational positions for each type of piece.

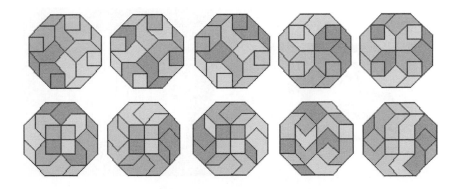

Figure 2. A puzzle that assembles into many patterns.

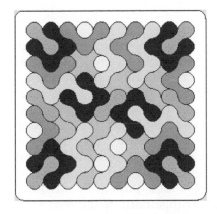

Figure 3. A very complex set of curvy shapes.

There is delight in such diversity and adaptability. The open-ended freedom of logically coherent groups, where more than one way is possible and viable, speaks to the deepest needs of one's being: my survival craves the assurance of many options; don't fence me in; don't run me into a *cul-de-sac*. Let the puzzle unfold in many ways, surprising in its diversity, its ability to embrace all its parts into loving union.

Gene Roddenberry (creator of *Star Trek*) had a motto: IDIC—infinite diversity from infinite combinations. He was speaking humanistically, of course. But it well describes the very process of creation.

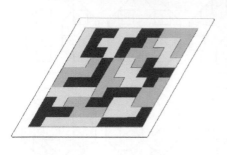 

Figure 4. Two tilings that combine multiple levels of conditions.

We can create ever more complex combinations. The two tilings in Figure 4 combine multiple levels of conditions. The one on the left has turned pentominoes into pentarhombs, and the rhombus-shaped array is tiled so no two of the same color share edges (with only three colors), and no two of the same pair (left- and right-hand versions of the same shape) are the same color. This set is also solvable so that all pieces of the same color are grouped into a single region—an extremely difficult solution to find by hand. The design on the right has 24 distinct directional, chevron-shaped tiles, each composed of four diamonds surrounding a vertex. Each tile has a different arrangement of the four colors, and the overall array contains 77 *quadrants* (2 × 2 clusters of diamonds) such that no two quadrants have the identical color pattern.

The more complex and restrictive the conditions, the fewer options remain, and they are generally harder to discover. While it is theoretically interesting to explore such rarefied combinations, in the scheme of *survival* mankind should seek to keep wide safety margins of multiple options. Happiness is proportional to such freedom. Survival in and of the universe is our grandest, the ultimate puzzle. I am content—no, ecstatic—to believe humanity is one of the superstrings in the universal continuum.

Solving puzzles like the two above can be a strenuous exercise in persistence and ingenuity. And we're dealing here with not more than 24 elements! Of course, one can simply appreciate the concept of such a problem without necessarily tackling it.

And finally, there is total delight in contemplating the meta-complexity of combinations of elements, from the Big Bang to the formation of galactic clusters, galaxies, solar systems, planetary systems, and our world—that enabled life to emerge, that eventually produced the complex organism that can perceive itself as existing, as being a part of all that is.

The energy matrix that defines an individual human being is a wondrous event, and the problem-solving faculty of its brain is its crown jewel. One of its prime directives is to grow, accumulate knowledge, survive. Another is to guard and nurture, preserve and expand its kind. I call this transcendental striving: Love.

Variations of the 14-15 Puzzle

Rodolfo Kurchan

The goal of the 14-15 problem is to go from position A to position B by sliding numbers to empty squares.

1	2	3	4
5	6	7	8
9	10	11	12
13	15	14	

Position A

1	2	3	4
5	6	7	8
9	10	11	12
13	14	15	

Position B

Everybody knows that there is no solution to this problem.

Here I present seven variations of the 14-15 Puzzle, all of which do have solutions.

1. *The 14-15 multiples exchange.* You go from position A to position B by exchanging numbers whose sum is a multiple of 14 or 15 (*cf.* the name of this puzzle).

Example:

1	2	3	4
5	**8**	7	**6**
9	10	11	12
13	15	14	

1	2	3	4
5	8	7	**9**
6	10	11	12
13	15	14	

You can exchange numbers 6 and 8 because $6 + 8 = 14$. Then you can exchange 6 and 9, for example, because $6 + 9 = 15$.

2. *Square exchanges.* You go from position A to position B by exchanging numbers whose sum is a square number.

   Example:

3	2	1	4
5	6	7	8
9	10	11	12
13	15	14	

13	2	1	4
5	6	7	8
9	10	11	12
3	15	14	

   You can exchange numbers 1 and 3 because $1 + 3 = 4$; then, for example, you can exchange 3 and 13, because $3 + 13 = 16$.

3. *Neighbors sum.* Go from position A to position B by moving numbers to an empty space, subject to the rule that you cannot move a number if the sum of itself and its neighbors in the same row or column (not diagonal) is 40 or more.

   Example:

	2	3	4
5	6	7	8
9	10	11	12
13	15	14	1

1	2	3	4
5	6	7	8
9	10	11	12
13		14	15

   You can move the number 1 because the sum $1 + 12 + 14 = 27$ is less than 40, but you cannot move 15 because $15 + 14 + 12 = 41$.

4. *Neighbors multiples.* In this case you can move a number as long as the sum of itself and its neighbors in the same row or column (not diagonal) is:

   (a) a multiple of 4.

   (b) (Is it possible to replace multiples of 4 by multiples of 3?)

   Example: This is the solution for multiples of 2:

1	2	3	4
5	6	7	8
9	10	11	12
13	15		**14**

1	2	3	4
5	6	7	8
9	10	11	12
13		**15**	14

(1) 14 (12 + 14 = 26)   (2) 15 (15 + 11 + 14 = 40)

1	2	3	4
5	6	7	8
9	10	11	12
13	**14**	15	

(3) 14(10 + 13 + 14 + 15 = 52)

Total = 3 moves

5. *Strip.* Moving any of the four rows or four columns (they move as a continuous line with no end), go from position A to position B with the smallest possible number of moves.

Example:

	E	F	G	H
A	1	2	3	4
B	5	6	7	8
C	9	10	11	12
D	13	15	14	

	E	F	G	H
A	1	2	3	4
B	6	7	8	5
C	9	10	11	12
D	13	15	14	

B 3 = means move row B 3 times to the right.

6. *Moving numbers.* A counter can move (horizontally or vertically, but not diagonally) a number of squares equal to itself, so you can only move numbers 1, 2, or 3 (with 4 you would leave the board) and exchange positions with another counter.

How many moves do you need to go from position A to position B?

Example:

1	**4**	3	**2**
5	6	7	8
9	10	11	12
13	15	14	

5	4	3	2
1	6	7	8
9	10	11	12
13	15	14	

Example: Number 2 moves to the right and exchanges with 4.

Number 1 moves down and exchanges with 5.

7. *Jumping.* Counter movement is implemented by jumping over one or two counters (horizontally, vertically, or diagonally) to an empty space.

How many moves do you need to go from position A to position B?

Example:

1	2	3	
5	6	7	8
9	10	11	12
13	15	14	**4**

1		3	**2**
5	6	7	8
9	10	11	12
13	15	14	4

Number 4 jumps over 8 and 12.   Number 2 jumps over 3.

Open Problems

You can extend these problems to 5×5 or other boards.

Solutions

1. *The 14-15 multiples exchange:*

   14/1, 13/1, 13/15, 13/1, 14/1 = 5 moves

2. *Square exchanges:*

   14/11, 11/5, 5/4, 4/12, 12/13, 13/3, 3/1, 15/1, 1/3, 3/13, 13/12, 12/4, 4/5, 5/11, 11/14 = 15 moves

3. *Neighbors sum:*

   13, 14, 15, 14, 13 = 5 moves

4. *Neighbors multiples:*

   (a) 2, 14, 15, 14, 2 = 5 moves

1	2		4
5	6	7	8
9	10	11	12
13	14	15	3

   (b) Near solution: 13, 15, 14, 4, 13, 8, 4, 15, 13, 8, 3, 4, 8, 4 = 14 moves (only 3 is not in position)

5. *Strip:*

G1, D2, G3, D1, G1, D1, G3, D1 = 8 moves

6. *Moving numbers:*

3/14, 1/2, 1/14, 3/1, 1/15, 2/3, 3/13, 1/3, 3/14, 2/11, 2/9, 2/13, 1/13, 2/1, 2/9, 2/11, 2/1, 1/3, 1/2 = 19 moves

7. *Jumping:*

8, 15, 8, 13, 5, 15, 13, 5, 14, 15, 14, 5, 13, 14, 5, 13, 8, 14, 8 = 19 moves

Pentomino Battleships

Mogens Esrom Larsen

In the 1960s I first learned about pentominoes[1] from Martin Gardner [1, Chapter 13].Especially the figures on pages 119–120 were
appealing: playing with them and trying to occupy most of a chessboard was great fun. In order to explain the fun, we need to give
them names. Traditionally they are named by capital letters as
shown in Figure 1

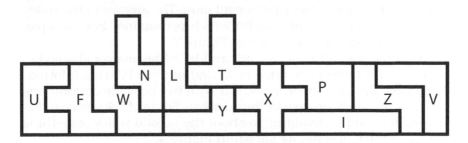

Figure 1. The 12 pentominoes.

---

[1]"Pentomino" is a registered trademark of Solomon W. Golumb [3].

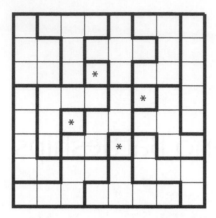

Figure 2. The 12 pentominoes occupying most of a chessboard.

The figures in Figure 72 [1] were inspiring: Look at the first, which is pictured in Figure 2

Battleships

Filling most of a chessboard with 64 squares with the 12 pentominoes covering 60 of them leaves 4 empty squares. Now, we made this into a variation of the game of battleships. We shot at the chessboard with a volley of four shots aiming at hitting the empty squares. Of course, we did usually hit some of the pentominoes. But after maybe four volleys we were able to locate the empty squares and shoot a volley hitting all four. The answer to the volley would be how many shots did hit which pentomino. For example, let's shoot in the picture shown in Figure 2.

We would say, "I hit a1, b1, a8, and h1." The answer should be, "you have hit twice in N, once in Z, and once in P." This information allows the location of the N, but with two possible orientations and that the other two are in the corners. So, to obtain more information, it would be tempting to shoot the second volley something like b2, h2, h7, and h8, as shown in Figure 4.

Now we learn that we have hit twice in I, once in L, and a second time in Z. This information allows us to place the N, the I, and the Z, but the P still has six possible orientations. And the L has five ways to orient. What we know for certain is shown in Figure 5.

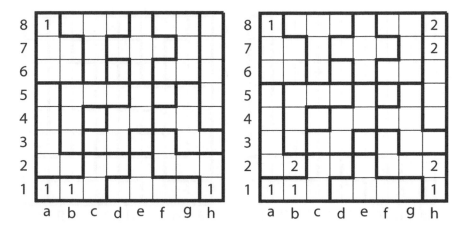

Figure 3. The first volley, with 1 marking the hits.

Figure 4. The second volley.

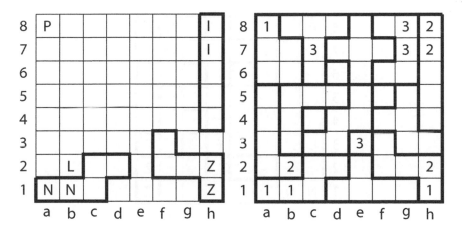

Figure 5. What we know after two volleys.

Figure 6. The third volley.

The next volley may be used to locate some missing orientation. For example, you shoot at e3 for a possibility for the L and possibly to show that the Y is between the N and the Z, c7 for a possible place of the P, and g7 and g8 to find another piece. Our hits are shown in Figure 6.

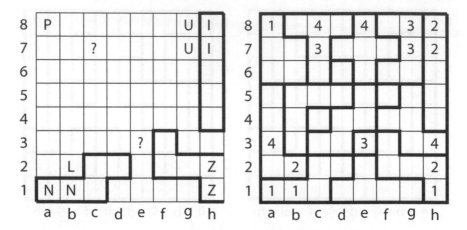

Figure 7. What we know after three volleys.

Figure 8. The fourth volley.

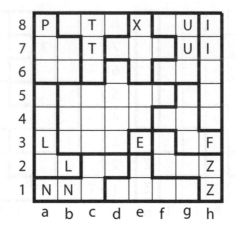

Figure 9. What we know after four volleys.

This time we learned that the U was hit twice; the T and an empty were hit once. So, we may reach a few conclusions—P has five ways left, the U has four ways, either the Y is as shaped between the N and the Z, or the T is located here with g1 as empty. And the L has only four ways. Not much! And the empty square may be one of two. Figure 7 gives a summary of what we know so far.

Now it is confusing, and we may try a3, c8, e8, and h3 for our next volley, shown in Figure 8.

We get the information that we have hit L, T, X, and F. Then T and E are located as c8 and e3, respectively. And L is placed, the F must be in h3. The U has two possible placements, but with the X in e8, there is only one. See Figure 9.

Only V and W are left, though V has only one place (the one we know) and W may be placed in two ways, leaving either c3 or e5 empty. So we must shoot a fifth time to be sure!

We have learned something inspiring. Although we have placed 11 figures, we still have to decide the exact location of the last, the W in the example. To make the game harder, we should look for such ambiguities.

Antisymmetry of the P

The P gives rise to the greatest ambiguity, placed in a 3×3 square it has 16 different ways to distribute the empty four! (Figure 10 shows two of them.)

Of course, it adds to the trouble to place this square as far from the corner as possible. So, we have the following problem: fill in the board with the 11 other pentominoes, leaving a square including most of the center to the P.

My five solutions are shown in Figure 11.

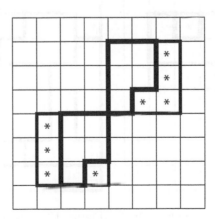

Figure 10. Two ways to place P in a 3×3 square.

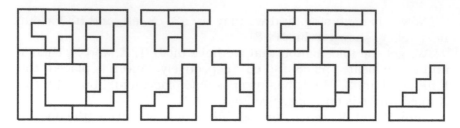

Figure 11. Five solutions for leaving a central square for P.

Alternative: Symmetry and Translation of the U

The U is symmetric, so turning it around only gives four different orientations. But, inside the 3×3 square is room for translation giving rise to four new different orientations, leaving us with eight different distributions of the four empty squares. (Figure 12 shows two of them.)

And for this piece I have been able to place the square including the center of the board only one way. See Figure 13.

If I am right in finding exactly one solution, the search for ambiguity has reached a case of uniqueness, and this pattern can be solved by these 11 pieces in one and only one way.

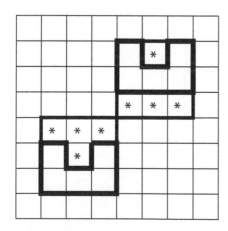

Figure 12. Two ways to place U in a 3×3 square.

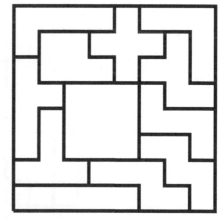

Figure 13. One solution for leaving a central square for U.

Bibliography

[1] Martin Gardner. *Mathematical Puzzles and Diversions from Scientific American.* London: G. Bell and Sons Ltd., 1961.

[2] Martin Gardner. *Morsom Matematik.* Copenhagen: Borgens Forlag, 1963.

[3] Solomon W. Golumb. *Polyominoes*, Second Edition. Princeton, NJ: Princeton University Press, 1994. (First edition published in 1965.)

[4] Mogens Esrom Larsen. "There Are No Holes Inside a Diamond." *Agate* 3 (1989), 30–36.

[5] George E. Martin. *Polyominoes: A Guide to Puzzles and Problems in Tiling.* Washington, DC: Mathematical Association of America, 1991.

Horses in the Stream and Other Short Stories

Earnest Hammingway

Error-correcting codes have become a key component of telecommunications in the half century since they were invented. Today, we depend heavily on them for everything from mobile phones and CDs/DVDs to internet security. Elementary linear codes can also form the basis for some two-person, mathematical magic effects. We explore four situations where cards are selected at random by a spectator, and put on display, whereupon you request a minor alteration of the chosen cards. Your accomplice, who has been trained in advance, now enters the room, surveys the scene, and quickly announces what was changed.

We begin by explaining the workings of a $(5,2)$ linear binary code in which 2-bit messages $[a,b]$ are encoded as 5-bit codeword $[a,b,a,b,a+b]$. Here a and b are either 0 or 1, and the rules of binary arithmetic apply, namely, $0+0=0$ and $0+1=1$, whereas $1+1=0$. The three new bits appended to the original message M are sometimes referred to as *parity check digits*. This code is simpler—and less efficient—than those usually considered to be on the ground floor of the modern edifice of error-correcting codes, but we explain it in some detail and offer several magical applications.

There are $2^2 = 4$ possible messages, coded as follows:

message	codeword
00	00000
01	01011
10	10101
11	11110

If these 2-bit messages are transmitted, via the 5-bit codewords listed above, and possible errors introduced—meaning that some of the bits get flipped from a 0 to a 1, or vice versa—then any of $2^5 = 32$ messages could be received, only 4 of which are uncorrupted.

This set of 32 messages forms a 5-dimensional vector space over the field $\{0,1\}$, and is equipped with a metric, or distance function. The distance between two vectors is simply the number of bits (i.e., places) in which they disagree. The weight of a vector is its distance from 00000, namely, how many of its bits are 1.

As the table above reveals, among the (three) non-zero legitimate codewords, the minimum weight is 3. This is also the distance between any two of those three codewords. What is important about this linear code is that it *detects, and enables the correction of, single errors*, as we now demonstrate.

Of the 32 possible received messages, there are 8 with double errors, but we'll focus on the $32 - 4 - 8 = 20$ corrupted messages R arising from the 5 different single-error transmissions for each of the 4 legitimate messages M. These are shown—together with the weights of the received messages R—in the first three columns of Table 1.

Given a received message R, we must determine the original message M.

A study of Table 1 reveals that if the weight of R is 1, then $M = 00000$, whereas if the weight of R is 3 or 5, then $M = 11110$. While the other half of the cases are not so obvious at first glance, these "weighty" observations are worth bearing in mind when performing associated card tricks.

Let's take another approach, which turns out to be easier and more logical (and also proves useful in the extensions to larger codes considered in [1]). The fourth column of Table 1 gives the "pretend" (and consistent) codewords P that are obtained from the first two bits of the erroneous Rs. In other words, the third, fourth, and fifth entries of each P are the parity check digits $[x, y, x + y]$

Correct codeword M	Corrupted codeword R	Weight of R	P based on corruption R	Where R, P disagree
00000	10000	1	10101	$x, x+y$
00000	01000	1	01011	$y, x+y$
00000	00100	1	00000	x
00000	00010	1	00000	y
00000	00001	1	00000	$x+y$
01011	11011	4	11110	$x, x+y$
01011	00011	2	00000	$y, x+y$
01011	01111	4	01011	x
01011	01001	2	01011	y
01011	01010	4	01011	$x+y$
10101	00101	2	00000	$x, x+y$
10101	11101	4	11110	$y, x+y$
10101	10001	2	10101	x
10101	10111	4	10101	y
10101	10100	2	10101	$x+y$
11110	01110	3	01011	$x, x+y$
11110	10110	3	10101	$y, x+y$
11110	11010	3	11110	x
11110	11100	3	11110	y
11110	11111	5	11110	$x+y$

Table 1. The 20 corrupted codewords.

of the first two bits $[x, y]$ of the corresponding R. Finally, the last column of Table 1 identifies the exact locations in which P and R differ—namely which of the parity check digits fail to agree. These are listed in terms of their constituent entries: $[x, y, x+y]$. Now a clear pattern emerges, leading to the following algorithm.

Algorithm 1 (Decoding algorithm for (5,2) code).

Given a codeword R, known to contain exactly one error, use its first two bits to generate and append "pretend" parity check digits, thus forming a new self-consistent codeword P. See where R and P differ:

If it's in the third and fifth positions, the error in R is in the first bit.

If it's in the fourth and fifth positions, the error in R is in the second bit.

If it's in the third position only, the error in R is in the third bit.

If it's in the fourth position only, the error in R is in the fourth bit.

If it's in the fifth position only, the error in R is in the fifth bit.

Of course, if the weight of R is odd, we can always fall back on our earlier observation to decode rapidly. Or readers may prefer to devise their own (perhaps simpler) algorithms.

At last, we are ready to have some fun applying this to card magic. We could work with just five card positions, but it's really no harder to use six, and it seems more fair in our first routine.

Horses in the Stream

The audience member is invited to select any three cards from the deck, and lay them in a face-up row on the table. You supplement this row with three more face-up cards of your own choosing. "Think of these cards as six horses, in the stream here!" With a flourish, indicate an imaginary stream. Before that sinks in, add, "No doubt you've heard the expression, 'Don't change horses in the middle of a stream.' Actually, that's exactly what I want you to do now. Please change any one horse—for a horse of a different color!"

The audience member replaces any one of the cards in the row with a new card from the deck, with the proviso that the new card must not be the same color as the one it is substituted for. Your accomplice now enters the room for the first time, and soon identifies which card in the row was switched.

Here's how it works. Your three choices are based on the colors of their first two; their third card color (denoted by X below) is ignored. Use the convention that 0 represents black, and 1 represents red, and code $[a, b, x]$ as $[a, b, x, a, b, a + b]$. Hence: If they pick BBX, you pick BBB. If they pick BRX, you pick BRR. If they pick RBX, you pick RBR. If they pick RRX, you pick RRB. When the switch is made, clearly it's the third card if and only if the others determine a legal codeword (i.e., are self-consistent). Otherwise, one of the 20 cases in Table 1 arises (for the five positions/cards omitting the third one). The decoding above can be used to determine which color sequence was in place before the switch, and hence which card is the switched one.

Figure 1. Consistently correct.

Figure 2. Stop the deliberate error.

For example, suppose the audience member picks 10♣, Q♦, 2♡, and you—concerned only with the colors of their first two selections—pick J♠, 5♡, 7♦. Before the switch everybody sees the cards as shown in Figure 1.

Now let's assume the spectator discards the second card, Q♦, and in its place puts 8♣. When your accomplice enters the room, he sees the cards as depicted in Figure 2.

Mentally, he ignores the third card for now, and converts the other five colors to $R = 00011$. He notices that this is not self-consistent, so the ignored third card is not the problem. He has also just worked out $P = 00000$, which differs from R in the x and $x + y$ positions. Hence, he knows that the error in R is in the second bit, and so the 8♣ must be the switched card.

The idea of applying Hamming codes like this to two-person card tricks dates back to 2002, and is due to mathematician Jeffrey Ehme of Spelman College. He may have been inspired, in part, by some wonderful creations of Alex Elmsley's (e.g., "The Octal Pencil" [2, page 93]).

Another version of this can be performed where the cards are chosen alternately. First the spectator picks one, then you pick a matching one, and so on, resulting in $[a, a, b, b, x, a + b]$.

If, in either version, the audience member picks three black cards, then as you pick three more, it may come as no surprise to onlookers that your accomplice is able to spot one red card among five black ones! However, this should only happen one time out of eight, on average. Moreover, your accomplice would have no reason to suspect an initial monochrome arrangement.

We could, of course, use a less obvious concept than color as our binary identifier—perhaps odd/even or prime/composite values—but this creates its own difficulties. If you have to explain your new, improved, and oh-so-subtle deck division scheme, just to force the spectator to do the correct kind of card switch, that more or less defeats the whole point of it!

One way around this is to divert attention from the card faces from the very start, and perform most of the trick with the cards face-down.

Men Without Faces

Take out a deck of cards, and shuffle it. Split the deck into two halves, and put these on the table, face down. Invite an audience member to pick out any card, from either packet, and place it face down on the table, to one side. Repeat twice, so that there are three face-down cards in a row on the table in addition to the two packets. Next, *you* select three more cards from the packets and place them beside the first three, choosing "at random" between the two packets. There is now a row of six cards on the table. Stress that nobody could possibly know what any of them are— which is true! Reform the deck (of $52 - 6 = 48$ cards) by placing one packet on top of the other. Turn away.

Have the audience member hand you any one of the cards on the table, which you replace in the deck. Next, invite him to choose a different card from the deck. Have him note its face, and show it around for everybody except you to see. Finally, have this new card replaced in the gap in the row, face down, so that once again there are six cards on the table.

Turn back, and have the cards flipped over so that (for the first time) all faces are visible. Recall the fairness of the procedure as you shuffle the rest of the deck: cards were randomly picked— three by the audience member, three by you—without anybody knowing what any of them were. The audience member then decided which card would be set aside, and which one would replace it. You yourself did not witness the switch, and you no have idea which card on the table was shown around to the audience.

Say, "You'd be amazed if I could now tell you which card was switched—and I certainly could, after all my years performing magic.[1] But I have a definite advantage; I've been holding the cards in my hands for a few minutes, absorbing their distinctive vibrations." Continue, "Let's see if *somebody who has not seen or heard any of what has happened* can do a little magic!"

Your accomplice enters the room for the first time, and after a moment's reflection, correctly picks out the replacement card from the row of six.

Here is one way to do this. First split the deck into red and black halves, and place these together, say with the blacks on top. Some casual in-hand shuffling—of both halves of the deck separately, keeping the blacks on top—will be fairly convincing.

The audience member picks three cards completely randomly, and all you have to do is pay attention to which pile her first two selections come from. As in the previous trick, you match her first two choices, colorwise, and for your final selection you pick according to the "sum" of the colors (i.e., pick black if and only if the first two cards are the same color).

Now we move on to the switch: when you turn away and are handed one card, peek at it and note its color. *It is essential that the replaced card be of the opposite color—this determines which part of the deck you offer for that selection!* Since you shuffle— properly, this time—before the end of the trick, the evidence is destroyed even if the deck is inspected later.

The rest of the trick is identical to the last one: as soon as the cards are turned face up your accomplice sees everything and decodes based on colors.

Your claim that you could do the trick unaided is based on the fact that after the switch is done, with your back turned, you know the color of the card in question, *and*—since you know the colors of all of the original cards too—you can narrow down the choices for its position. By peeking at some of the face-down card faces before your accomplice comes in, and muttering, "As I suspected," you can even scribble down a prediction that can later be checked.

In this incarnation of the basic trick, it's not such a bad idea to try a less obvious deck division, such as prime/composite values— just throw caution to the wind and lump the aces in with one group! That way you can even flash card faces early on, such as

---

[1] There is an element of truth to this claim; keep reading.

when asking for selections to be made, boldly emphasizing, "Pick a red or a black card, your choice."

Offering the entire deck for the selection of the substitution card is possible, provided that you force this card to come from the half you want. It's true that a similar trick can be done—even without any coding theory—in which *all* selections can (appear to) be from a full deck, if you are good at forcing, but this takes us further and further from mathematical principles!

The basic five-card code, $[a, b, a, b, a + b]$, is unique up to permutation of the five bits, since a, b, and $a + b$ are the only non-trivial linear combinations of a and b (mod 2). We've been assuming that a, b were freely chosen by the spectator, and then the three check digits were forced. If the trick is to be done for the same audience several times, it might be better to use a different arrangement, such as $[a, b, a, a + b, b]$. Of course if we do the six card version originally suggested, we can use *any two* of the first three cards as the $[a, b]$ that are encoded.

The Nth Column

Since it was Fitch Cheney's classic "Given any five cards" card trick plot from the late 1940s that first inspired us to explore mathematical card tricks in some depth, it seems fitting to include a trick in a similar spirit here. It hinges on the following result, whose proof is given in due course.

> If we choose any five cards at random from a deck, then it is possible to arrange them in a row in such a way that if one card is exchanged for a new one of a different color, then the position (and hence identity) of the new card can be detected by somebody who only sees the end result.

Proceed as follows. Hand the deck to a spectator, and say, "Shuffle the cards well, and deal out some poker hands. Pick one of those hands for yourself, and put the rest aside; we won't need them. Let me have a look." You take the five cards and move them around, commenting briefly on the spectator's luck and prospects.

"I'm going to make you an offer you don't usually get in real life," you say as you spread the cards in a neat face-up row. "You may like this hand, you may not. Frankly, it could be better. I'm going to give you the opportunity to replace any one of these cards with a better one! You choose which one you want to get rid of. Take your time—you only get to do this once."

The spectator points to a card, and you slide it out. "I promised you a better card, did I not? You pick one—you can use any card at all from the rest of the deck, provided it's the opposite color of the one you just discarded." Have the spectator slide the new card into the gap in the row. The discarded card is set aside.

Call in your accomplice from the next room. "My friend here has studied human psychology a lot, and he reckons he can figure out which card you switched by asking you to read out the card names in any order you wish and listening to your voice modulations. He says that when you get to the card in question, you'll get emotional. Be careful, he's yet to make a mistake." Your accomplice delivers!

Here is the basic idea: arrange the five cards in a way that they are correctly coded, regardless of what selection of black and red cards we have. Bear in mind that the four possible codings considered in our very first table give rise (assuming 0 = black and 1 = red) to rows of five cards with no reds, three reds, or four reds.

So, if you are handed five cards with one of those possible numbers of reds, you are home and dry—you even have two ways to arrange the cards in the three reds case. But what if you receive one red, two reds, or five reds? Then you have four, three, or zero black cards, since $5 - \{1,2,5\} = \{4,3,0\}$. Simply change the convention: now 0 = red and 1 = black!

Fortunately, this can be done in such a way that your accomplice *knows* if there has been a convention switch. There are six cases to consider, three good and three not-so-bad:

(0) You've zero red cards. Proceed as usual (but see below).

(1) You've one red card. Indicate a convention switch and mimic case 4.

(2) You've two red cards. Indicate a convention switch and mimic case 3.

(3) You've three red cards. Proceed as usual (but see below).

(4) You've four red cards. Proceed as usual (but see below).

(5) You've five red cards. Indicate a convention switch and mimic case 0.

Before we spell this out in detail, we must explain the method of indicating the convention switch. So far we've only paid attention to the card colors, not their values. Put a total linear ordering on the whole deck, such as: A♣, 2♣, ..., K♣, A♡, ..., K♡, A♠, ..., K♠, A♢, ..., K♢. With this understood, we are at liberty to use the values of the red *and* the black cards to communicate further information.

In any row of five cards, there will be "subrows" of both red and black cards. (If we have all one color, of course there is no subrow of the other color, and if we have four of one color, the other subrow consists of just one card!) If we intend the usual convention to be understood (namely, 0 = black), which is to say cases 0, 3, or 4 above, we can arrange the cards coded so that any visible subrows are rising, with respect to the total ordering on the deck. On the other hand, if we intend the convention to be switched (namely, 0 = red), as in cases 1, 2, or 5 above, we can arrange it so that all subrows are falling.

Since the rising/falling considerations apply to just two suits (of the same color) at a time, it's not so hard for either party involved to "deal with" the ordering considerations quickly.

The point is that when a card is exchanged later, which might mess up a rising or falling subrow, all is not lost. If the affected subrow is of length four or five—after the exchange—then a single card out of sequence still leaves no doubt as to the original rising/falling status.

If, on the other hand, after the exchange we have subrows of length two and three, one of three situations is in effect: one subrow is rising and the other is falling (in which case the one of length three is the "correct" one), both are monotone *in the same sense*, or only one (of length two) is monotone. These last cases will be established in detail below.

It's time to try all of this in practice. We present the three easy cases first.

- Case 0. You've zero red cards. So your cards are B_1, B_2, B_3, B_4, B_5, in ascending order with respect to the agreed-upon total ordering of the whole deck. You arrange them in that order: after one exchange, your accomplice sees (a rising subrow of) four blacks, knows the usual convention applies, and decodes to 00000 = *BBBBB* (i.e., knows that the lone red was the exchanged card) as required.

- Case 3. You've three red cards. So your cards are B_1, B_2, R_1, R_2, R_3, each subrow in ascending order. You rearrange them in the order B_1, R_1, B_2, R_2, R_3. Without loss of generality, the cards are 2♣, 2♡, 4♣, 4♡, 6♡.

Now consider the possibilities after one exchange. First let's assume a black card is removed and replaced by a red card, there are four possible relative values for the red card—A♡, 3♡, 5♡, or 7♡—and no matter which of the two black cards it replaces, the new subrow of four reds is either rising or rising-dominant. Now we assume a red card is removed and replaced by a black one, say A♣, 3♣, or 5♣. No matter which of the three reds it replaces, the remaining subrow of two reds is rising (and the new subrow of three blacks is either rising or non-monotone). In all subcases, after one exchange your accomplice knows that the usual convention applies, and decodes to 01011 = *BRBRR* as required.

- Case 4. You've four red cards. So your cards are R_1, R_2, R_3, R_4, B, which you arrange in that order. Without loss of generality, the cards are 2♡, 4♡, 6♡, 8♡, and 2♣.

  Now consider the possibilities after one exchange. First let's assume the black card is removed and replaced by a red card, there are five possible relative values for the red card—A♡, 3♡, 5♡, 7♡, or 9♡—and the new row of five reds is again either rising or rising-dominant. Now we assume a red card is removed and replaced by a black one, say A♣ or 3♣. In the latter case, a bogus falling subrow of length two is created, but we still have a rising subrow of length three to guide the way. In the former case, we have two rising subrows. In any case, your accomplice knows that the usual convention applies, and decodes to 11110 = *RRRRB* as required.

It only remains to do all of this "reversed" for the other three cases, ensuring that we can communicate the convention switch.

- Case 1. You've one red card. So your cards are B_1, B_2, B_3, B_4, R. You arrange them B_4, B_3, B_2, B_1, R—after one exchange, your accomplice sees a falling-dominant subrow of (at least) three blacks, and regardless of the two reds which may be present, knows that a switch in convention applies, decoding to $00001 = BBBBR$ as required.

- Case 2. You've two red cards. So your cards are R_1, R_2, B_1, B_2, B_3, each subrow in ascending order. You rearrange them in the order R_2, B_3, R_1, B_2, B_1. Without loss of generality, the cards are 4♡, 6♣, 2♡, 4♣ and 2♣.

  Now consider the possibilities after one exchange. First let's assume a red card is removed and replaced by a black card, without loss of generality it's A♣, 3♣, 5♣, or 7♣, and no matter which of the two reds it replaces, the new subrow of four blacks is either falling or falling-dominant. Now we assume a black card is removed and replaced by a red, say A♡, 3♡, or 5♡. No matter which of the three blacks it replaces, the remaining subrow of two blacks is falling (and the new subrow of three reds is either falling or non-monotone). In all subcases, after one exchange your accomplice knows that a switch in convention applies, and decodes to $01011 = RBRBB$ as required.

- Case 5. You've five red cards. So your cards are R_1, R_2, R_3, R_4, R_5, in ascending order with respect to the agreed-upon total ordering of the whole deck. You arrange them in *reverse* order: after one exchange, your accomplice sees (a falling subrow of) four reds, knows the convention is switched, and decodes to $00000 = RRRRR$ (i.e., knows that the lone black was the exchanged card) as required.

Now that we are considering more than just the *colors* of the card faces, let's be a little more ambitious.

The general idea, so far, has been to identify the new card in a row, based on the fact that it differs in color—or some other key way—from the card it replaces, which has been set aside. Furthermore, we either explicitly requested that the two cards in question be of differing colors, or forced that to be true by only offering half the deck for the switch.

Wouldn't it be nice not to have to make any such request, offer the entire remainder of the deck for the switch, *and have our accomplice not only pick out the new card, but also the one it replaced?!* That is precisely what our final routine attempts to accomplish.

Let's rephrase the idea behind the usual Hamming tricks discussed earlier: by using a single binary identifier, such as color, we can code the cards in such a way that after a single *significant* (i.e., change of color) switch, our accomplice can both spot the new card and narrow down the identity of the one it replaces to (at most) half the deck. Some coaching is required at the switching stage—left totally to chance, it will only work about half of the time.

The basic idea that we explore next can be summarized this way: by coding in parallel, using three (forgive the confusing geometry) "mutually orthogonal" binary identifiers at the same time, we can guarantee that our accomplice will spot the new card and also narrow down the identity of the one it replaces to (at most) an eighth of the deck. We'll do all of this as painlessly as possible, while simultaneously eliminating the one chance in eight of failure, so that your accomplice nails the replaced card exactly!

To Have and To Have Not

The spectator picks three cards and you pick three, these are laid in a row. You ask for any one card to be selected and set aside, to be replaced presently with another one. You then run through the rest of the deck seeking your lucky card, which you set face down on the table.

The cards are shuffled, and spread (face up or face down) for the spectator to pick a replacement card. This is inserted in the gap left by the rejected card, which is itself inserted randomly in the deck. The cards are shuffled again. Your accomplice enters, passes your lucky card (face down) over the backs of the row on the table, and soon identifies the new card. Five times out of six, he can also pick the original rejected card out of the deck!

The key this time is that the three cards you chose must simultaneously code three characteristics of the first two of the spectator's cards: color (X), status (Y), and parity (Z). We associate a binary triple $[X, Y, Z]$ to each card, where X is 0 if and only if the card in question is black; Y is 0 if and only if the card value is low, which we take to be 2, 3, 4, 5, 6, or 7 (aces are considered high here); and Z is 0 if and only if the card value is 2, 4, 6, 8, 10, or Q (we also consider aces to be odd).

For instance, if the spectator picks Q♠, 4◇ and A♣, this corresponds to the triples $[0, 1, 0]$ and $[1, 0, 0]$, ignoring the third card

Figure 3. Can a card be switched out and remain anonymous?

as usual. Computing the binary check digits $[a, b, a + b]$ three times over, where $[a, b]$ is the X, Y, and Z entries of the spectator's cards respectively (namely, $[0, 1]$, $[1, 0]$, and $[0, 0]$), we find that your cards must correspond to the triples $[0, 1, 0]$, $[1, 0, 0]$, and $[1, 1, 0]$, respectively. That is to say, your first card must also be black, high, and even (which gives you five choices), your second card must also be red, low, and even (which gives you six choices), whereas your third card must be red, high, and even (which gives you six choices). Let's assume that you select 8♠, 6♡, and 10♢, as depicted in Figure 3.

As soon as the spectator indicates which card she intends to get rid of, you must run through the deck, under the pretext of finding your lucky card, and find the cards you do not want her to pick as the replacement. For now assume that she is not dumping the third card. For instance, if she indicates that she wishes to jettison 6♡ from Figure 3, you must ensure that she doesn't pick another red, low, and even card; as that would result in all three error checks (with respect to the different binary codings) failing for your accomplice. Hence, you find and cut to the top and/or bottom of the deck these cards: 2♢, 4♢, 6♢, 2♡, and 4♡. If you wish, set one of these aside face down as your so-called lucky card.

Under these circumstances, you can riffle shuffle, have a card selected from (almost) any part of the deck, and have the rejected card inserted somewhere in the deck before thoroughly losing it with further riffle shuffles that retain the top and bottom stock, thus setting the stage for a successful conclusion.

The replacing card will differ *in at least one regard*—color, status or parity—from the one it replaces. As a result, when your accomplice enters the room and does error checking and decoding—three separate times, one each for the various binary characteristics—he will find at least one instance of error, and hence be able to pinpoint the position of the card switch. Better yet, by keeping track of all errors that occur, he will know the precise characteristics of the dumped card. Since you obligingly collected all matching cards (*except* the correct one) at the top and bottom of the deck earlier, only one card *within* the deck will fit the bill. This is the one your accomplice picks out, perhaps with the aid of your lucky card; but it must be clear that the face of this card is never seen!

On the other hand, if the third card—which played no role in your heavily coded choices—is the one selected for dumping, you are out of luck. Your accomplice will see that all error checking points to total consistency, and hence will know that the third was the switched card. Then that's it: just what card was discarded and shuffled into the deck will remain a mystery, unless one uses the so-called lucky card to convey more information.

One option is to rework everything using a more advanced, $(6,3)$ linear binary code, or you could just settle for a 5/6 success rate. If you have not announced ahead of time exactly what you intend to accomplish—always a good idea in magic—nobody will even know that you were hoping for more than you were able to deliver.

Bibliography

[1] Jeffrey Ehme and Colm Mulcahy. "Hamming It Up With a Deck of Cards." To appear.

[2] Stephen Minch. *The Collected Works of Alex Elmsley*, Vol. 2. Tahoma, CA: L&L Publishing, 1994.

A Potpourri of Polygonal and Polyhedral Puzzles: A Collection of Combinatorial Coincidences

Alan H. Schoen

Described below are several tiling and packing puzzles derived from orderly arrays of pairs of positive integers. Some of these puzzles are geometrical expressions of the *Triangular Array*, $T(n)$, shown in Figure 1(a), which is composed of the $\binom{n}{2}$ ordered pairs of row and column indices from 2 to n. Others are based on the $(n-1)^2$ entries in the *Square Array*, $S(n)$, of Figure 1(b).

(2,2)(2,3)	$\cdots$	(2,n)		(2,2)(2,3)	$\cdots$	(2,n)
(3,3)	$\cdots$	(3,n)		(3,2)(3,3)	$\cdots$	(3,n)
$\cdots$					$\cdots$	
		(n,n)		($n,2$)($n,3$)	$\cdots$	(n,n)
	(a)				(b)	

Figure 1. (a) $T(n)$. (b) $S(n)$.

Lominoes

Ln, the *standard set* of lominoes of order n, contains one specimen of every L-shaped polyomino with arms of unit cross-section area that can be cut from a $1 \times n \times n$ grid of cubes. The number $N(n)$ of lominoes in Ln is $\binom{n}{2}$ and their total volume $V(n)$ is $(n+1)\binom{n}{2}$. The most versatile sets are the 28-piece standard set L8 (see Figure 2) and the 32-piece *augmented set* L8†, which consists of L8 plus a duplicate of each of the pieces, shown in black, that lie in the NW/SE diagonal strip of Figure 2.

Every lomino is named for the ordered pair of indices (i, j) (see Figure 1(a)) that specify the *armlengths* of its two arms. Since every lomino may be turned over, it is named $[i, j]$ in one orientation and $[j, i]$ in the other. If it is oriented like the letter L, with a *stem* of length i and a *base* of length j, we name it $[i, j]$. Figure 4 shows the two different forms of the piece that correspond to [5,7]. Each of these shapes is called the *transpose* of the other. A lomino is called *face up* if $i < j$, *face down* if $i > j$, and *face neutral* if $i = j$. A lomino is said to be facing NE, NW, SW, or SE, according to the direction in which its concave corner is facing. In Figure 4, both [5,7] and [7,5] face NE.

Lominoes whose name labels in Figure 3 lie in identically shaded NW-SE diagonal strips of squares belong to the same *pronic rectangular subset* (see Figure 5). Each such diagonal strip in Figure 3 is identified by the number of the subset to which its pieces belong. Every standard set of odd order and every augmented set of even order can be partitioned into $\lfloor n/2 \rfloor$ such $n \times (n+1)$ rectangles. For

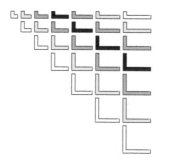

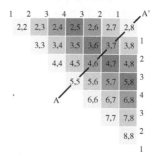

Figure 2. L8 arranged according to T(8).

Figure 3. Partition of L8 into four subsets.

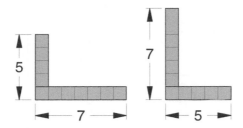

Figure 4. A lomino *face up* (left) and *face down* (right).

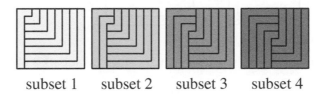

subset 1   subset 2   subset 3   subset 4

Figure 5. The four pronic rectangular subsets of L8†, canonically colored.

$k = 1$, 2, and 3, subset k of L8† contains the eight lominoes $[i, j]$ for which $j - i = k - 1$ or $j - i = 7 - k$. For $k = 4$, subset 4 of L8† contains two specimens of each of the four lominoes $[i, j]$ for which $j - i = k - 1 = 7 - k = 3$. A set is called *canonically colored* if each of its pronic rectangular subsets is of a different color, as in Figure 5. Any pair of pieces $[i, j]$ and $[n + 2 - j, n + 2 - i]$ that are related by reflection in the *medial line* AA′ (see Figure 3) are called *dual*.

A variety of exercises for L8, L8†, and the other puzzles introduced here are described below. Some lominoes exercises that may seem amenable only to brute force attack have simple solutions based on the orderly structure of $T(n)$.

Corrals and Fences

A *corral* is a self-avoiding circuit tiled by one pronic rectangular subset. Subsets 2 and 4 of L8† each admit many corral tilings, but it has been proved that subsets 1 and 3 admit none [11]. The dual of a corral is obtained by replacing every lomino by its dual, oriented face up if the lomino is face down (and vice versa) and facing in the same direction. If the CCW *signatures* of dual corrals like those of Figure 6 are compared, the dual relation be-

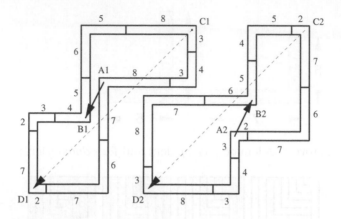

Figure 6. Two dual corrals—each is tiled by subset 2 of L8†.

tween the two lominoes of each corresponding pair is immediately
apparent.

$$\langle[7,2][7,6][8,7][3,4][3,8][5,6][4,5][3,2]\rangle \quad \text{(left corral)}$$
$$\updownarrow\updownarrow \quad \updownarrow\updownarrow\updownarrow\updownarrow \quad \updownarrow\updownarrow\updownarrow\updownarrow \quad \updownarrow\updownarrow\updownarrow\updownarrow \quad \updownarrow\updownarrow$$
$$\langle[3,8][3,4][2,3][7,6][7,2][5,4][6,5][7,8]\rangle \quad \text{(right corral)}.$$

Solutions for starred (*) problems are in the Solutions section
below. Solutions for unstarred problems appear in the CD manual,
Lominoes [11].

Exercise 1. Prove that the dual of a corral is a circuit (but not necessarily a *self-avoiding* circuit).

Exercise 2. Prove that for every pair of dual corrals,

(a) the directed lengths analogous to (A1→B1) and (A2→B2) in
    Figure 6 are equal and opposite, and

(b) the lengths analogous to |(C1→D1)| and |(C2→D2)| are equal.

Exercise 3. Prove that it is impossible to tile a corral with either
subset 1 or subset 3 of L8†. Hint: use a parity argument here and
in Exercise 6 [5, 11].

Exercise 4. Construct a pair of dual corrals, each tiled by subset 2 of L8†, that are *oppositely congruent*.

Exercise 5 ("Farm" problem [3]). A self-avoiding circuit composed of the lominoes of Ln or L$n^\dagger$ joined end-to-end is called a *fence* of order n. Tile a standard fence with the 28 lominoes of L8.

Exercise 6. A standard fence of order n is called *matched* if the two contiguous arms of every pair of adjacent lominoes have the same armlength. Prove that no matched FENCE exists for $n \neq 8k$ ($k = 1, 2, \ldots$) [11].

Conjecture 1. *For odd $n \geq 5$, there exists a simply-connected string, matched everywhere except at its ends, containing all but $(n-3)/2$ of the $\binom{n}{2}$ lominoes in Ln.*

Exercise 7. Tile a matched fence of order 8.

Sawtooths and Filigrees

A *sawtooth* of order n is a periodic zigzag composed of all the pieces of Ln or L$n^\dagger$ joined end-to-end in an alternating pattern of uniform slant height $H = n + 2$. A computer search reveals that there are 272 different sawtooths of order five. For $n > 5$, the number of sawtooths is unknown. A fragment of an L8 sawtooth is shown in Figure 7. A *filigree* is a standard sawtooth folded into a closed circuit with the symmetry of a square (see Figure 8).

Exercise 8. Prove that there exists a tiling of a standard sawtooth of order n for $n \geq 3$. (This conjecture was proved by Greg Martin in 1999 [8].)

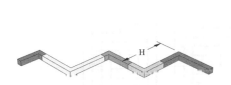

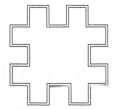

Figure 7. Fragment of L8 sawtooth. The slant height $H = 10$.

Figure 8. L8 filigree (folded sawtooth).

Exercise 9. Prove that a standard sawtooth of order n can be folded into a filigree only if $n = 8 + 32j$ or $n = 25 + 32j$ $(j = 1, 2, \ldots)$.

Squares, Rectangles, and Square Annuli

In contrast to the orderly tilings of $n \times (n+1)$ pronic rectangles, there are no known algorithms for tiling squares with pieces of Ln or L$n^\dagger$. It is not difficult, however, to tile a single square of any size from 5×5 to 15×15 with pieces of L8. Figure 9 shows a few examples. Tiling *two* same-size squares with pieces drawn from one L8 or L8† set, on the other hand, is a challenging exercise.

The holey rectangles in Figure 10, each of which is tiled by one L8 set, are tiling arenas inspired by a suggestion from Ed Pegg Jr. [9]. The arena on the left is 15×17; the one on the right is 11×23.

Exercise 10. Prove that it is impossible to tile a square smaller than 5×5 with pieces drawn from one L8 set.

Exercise 11. Tile both a 7×7 square and an 8×8 square with pieces drawn from one L8 set.

Exercise 12. Tile two 8×8 squares with pieces drawn from one L8 set.

Conjecture 2. *It is impossible to tile two same-size SQUARES smaller than 8×8 with pieces drawn from one L8 set.*

Exercise 13.

(a) Tile two 12×12 squares with the 32 pieces of L8†.

(b) Verify that for $n=8k$ $(k = 1, 2, 3, \ldots)$, the area of the augmented set L$n^\dagger$ is equal to that of two congruent squares [11].

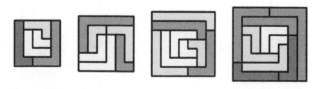

Figure 9. Tilings of 5×5, 6×6, 7×7, and 8×8 squares by lominoes of L8.

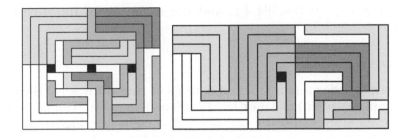

Figure 10. Two holey rectangles.

Exercise 14. Tile a 24×24 square with two $L8^\dagger$ sets. Is it possible for more than five intact pronic rectangular subsets to be embedded in the tiling?

Exercise 15. Tile the following centered *square annuli* with one set [11]: $8^2 - 2^2$(L5), $13^2 - 1^2$ (L7), $16^2 - 2^2$ (L8), $17^2 - 1^2$ ($L8^\dagger$), $18^2 - 6^2$ ($L8^\dagger$).

Exercise 16. Tile the following sets of RECTANGLES with one set [11]: one 14×18 (L8), two 9×14 (L8), three 8×12 ($L8^\dagger$).

z-RINGS

A square annulus tiled by z lominoes is called a *z-ring* $(2 \leq z \leq 4)$. The edge length w of the outer square boundary of a z-ring is called its *ringwidth*. A z-ring is defined by its CCW signature $\langle [i_1, j_1][i_2, j_2][i_3, j_3][i_4, j_4] \rangle$. For 2-rings and 3-rings, fictitious $[0, 1]$ pieces are inserted at appropriate positions in the signature to ensure that all four of the sums $j_k + i_{k+1}$ $(1 \leq k \leq 4, i_5 = i_1)$ are equal to the ringwidth. Hence the signature of the 3-ring in Fig-

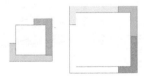

Figure 11. Two z-rings tiled by lominoes of L8.

ure 11 is $\langle [3,7][0,1][6,3][4,4] \rangle$, and the signature of the 4-ring is $\langle [8,3][8,8][3,6][5,3] \rangle$.

Towers and Ziggurats

A *tower* of order n is a stack of 4-rings of ringwidth $n+2$ tiled by all the pieces of either Ln or Ln^+ ($n \equiv 0$ or $1 \pmod 8$). The L8 tower has seven stories and the L8$^+$ tower has eight (see Figure 12). A *ziggurat* of order n is a stack of z-rings of consecutive ringwidths, tiled by Ln, forming a hollow stepped pyramid 13. A ziggurat is called *regular* if its z-rings are all 4-rings, and *irregular* otherwise. It is far harder to find a packing for the *regular* seven-story L8 ziggurat ($7 \leq w \leq 13$) and for the *irregular* nine-story L8 ziggurat ($4 \leq w \leq 12$) than it is to pack either of the towers of Figure 12.

We regard two z-rings composed of the same pieces as equivalent, even if they are differently arranged, or rotated, or reflected. For some 4-ring compositions, two or even three different arrangements are possible. Full tree searches reveal that there are 384 packing solutions (192 arrangements plus their duals) for the L8 regular ziggurat and 59 packing solutions for the irregular one.

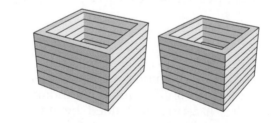

Figure 12. L8 and L8$^+$ towers.

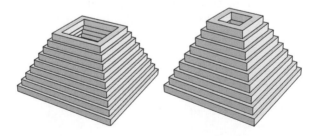

Figure 13. Two L8 ziggurats.

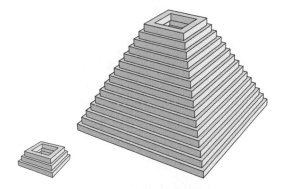

Figure 14. L5 and L11 ziggurats.

By contrast, a sparse Monte Carlo search for packings of the L8 and L8† towers found thousands of solutions. No algorithm exists for packing ziggurats, but there is one for packing towers of order $n \equiv 0$ or $1 \pmod 3$ (See Exercise 19).

It is not difficult to prove that (a) no ziggurat can be packed by an augmented lominoes set, and (b) aside from the two ziggurats in Figure 13, there are only three that can be packed by a standard lominoes set [11]:

(a) a 3-story irregular L5 ziggurat ($5 \leq w \leq 7$) (one solution);

(b) a 9-story regular L9 ziggurat ($7 \leq w \leq 15$) (6772 solutions);

(c) a 15-story irregular L11 ziggurat ($5 \leq w \leq 19$) (one solution has been found by hand [11]; the total number of solutions is unknown).

Exercise 17.

*(a) Find a packing for the L5 ziggurat and prove that it is unique.

(b) Find packings for the L8 and L8† towers of Figure 12.

(c) Find a packing for each of the two L8 ziggurats of Figure 13.

The volume $V_{\text{rings}}(a, b)$ of a set of z-rings with consecutive ring-widths w ($a \leq w \leq b$) is equal to $2(a + b - 2)(-a + b + 1)$. A necessary condition for the existence of a ziggurat packing is that $V(n) = V_{\text{rings}}(a, b)$, i.e.,

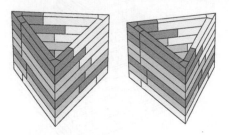

Figure 15. $_3$Tower packed by the set $_3$L7 of 21 $_3$lominoes.

$$n^3 - n - 4(a + b - 2)(-a + b + 1) = 0. \tag{1}$$

The diophantine Equation (1) has several other solutions besides those depicted in Figures 13–14, but I have proved that no packing exists for any of these other solutions [11].

Exercise 18. The *bending angle* at the interior corner of an ordinary lomino is $\pi/2$, but one can also define a more general shape called a $_p$lomino, with bending angle $2\pi/p$ ($p \geq 3$). If the prefix '$_p$' is omitted, it is implied that $p = 4$. Analogs of towers and ziggurats are defined for $p \geq 3$. The stereogram in Figure 15 shows a $_3$tower packed by the 21 $_3$lominoes of $_3$L7, a set that contains one $_3$lomino corresponding to each entry in $T(7)$ (see Figure 1(a)).

Let us define $_3M_{\mathrm{odd}}[w]$ and $_3M_{\mathrm{even}}[w]$ as the number of ways (ignoring rotations and reflections) to tile a trigonal 3-ring of odd and even ringwidths w, respectively. The ringwidth of a trigonal 3-ring is defined as the length of an outer horizontal edge.

Apply the Pólya-Burnside Lemma [2, 5, 11] to prove that

$$_3M_{\mathrm{odd}}[w] = (w - 2)(w - 3)(w - 4)/6, \tag{2}$$

and

$$_3M_{\mathrm{even}}[w] = (w - 2)(w - 3)(w - 4)/6 - (w - 4)/2. \tag{3}$$

Exercise 19. For $n \equiv 0$ or $1 \pmod 3$, exploit the symmetry of $T(n)$ to devise an algorithm for packing a $_3$tower with the $_3$lominoes of $_3$Ln [11].

Exercise 20. For $n \equiv 0$ or $1 \pmod 8$, devise an algorithm similar to that of Exercise 19 for packing a tower with the lominoes of Ln [11].

The total volume $V(n)$ of the lominoes in Ln is $n(n^2 - 1)/2$. Let us call the set $R(a, b)$ of z-rings of ringwidths $a, a + 1, \ldots, b - 1, b$ a *consecutive set*. One can prove that $a_{\min} = 4$ and $b_{\max} = 2n - 2$. Let us call the set $R(a_{\min}, b_{\max})$ of z-rings *maximal*. As stated above, $V_{\text{rings}}(a, b) = 2(a + b - 2)(-a + b + 1)$; hence the volume of a maximal set is $4n(2n - 5)$. For $n > 13$, it is impossible to squeeze an entire standard set into just one ziggurat, because $V(n) > 4n(2n - 5)$. (Note that $V(n) = O(n^3)$ and $4n(2n - 5) = O(n^2)$.) However, for $n = 8k$ and $n = 8k + 1$, the volume $V(n)$ is equal to that of k congruent regular ziggurats, and it is conjectured that packings exist for each of these cases.

The ringwidth w of a p-gonal ring is equal to the length of an outer horizontal edge of the ring. For $p \neq 4$, the number of $_p$lominoes in every p-gonal ring is necessarily equal to p. Full tree searches reveal that *no* packing exists for a five-story $_3$ziggurat by the 15 $_3$lominoes of $_3$L6 but that there are two dual packings of a seven-story $_3$ziggurat by the 21 $_3$lominoes of $_3$L7.

Conjecture 3. *Packings exist for every $_p$ziggurat for $p > 3$.*

Figure 16 shows a canonically colored packing of an 11-story $_5$ziggurat composed of the 55 $_5$lominoes of $_5$L11; $a = 8$ and $b = 18$. This is one of several solutions found in a Monte Carlo search, using Mathematica [11].

Exercise 21. Prove that for $n = 2p$ and $n = 2p + 1$ ($p \geq 3$), the number and volume of the pieces in $_p$Ln satisfy the necessary conditions for tiling the ρ p-gonal rings of one $_p$ziggurat, for which $a = \lfloor (n + 6)/2 \rfloor$, $b = \lfloor 3(n + 1)/2 \rfloor$, and $\rho = 2\lfloor (n + 1)/2 \rfloor - 1$ [11].

Quarks, Cache-22, and Related Puzzles

Every entry in $T(9)$ can be expressed geometrically not only by a $_3$lomino but also by a quadrangular *kite*. Figure 17(a) shows an equilateral triangle tiled by three kites with signature $\langle [4, 5][6, 8][3, 7] \rangle$, superimposed on a trigonal ring of $_3$lominoes. The area of a kite is proportional to its index sum. The kite vertex at the center of the associated trigonal ring is called *central*. The algorithm of

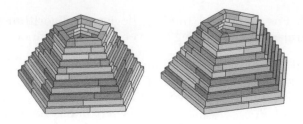

Figure 16.   Pentagonal $_5$ziggurat composed of the 55 $_5$lominoes of $_5$L11 (stereogram).

Exercise 19 defines those values of n for which the kite set $K(n)$, based on $T(n)$, can be arranged to tile a set of congruent equilateral triangles.

Figure 17 shows two six-color tilings of a hexagram by the 36 kites of $K(9)$ (*quarks*), distributed among 12 equilateral triangles of edge length 11. In the *packaged* tiling of Figure 17(b), the kites of each color are sequestered in two triangles. The object is to rearrange the kites so that every triangle is *pied*, i.e., tiled in three colors, as in Figure 17(c). With the color distribution of Figure 17(b), it is found that 65 of the 84 possible tilings of a single triangle (see Equation (2)) are pied.

Conjecture 4. *The only pied tilings of the hexagram are the 28 solutions found in repeated Monte Carlo searches, using Mathematica.*

Conjecture 5. *Every pied tiling can be arranged as a map-colored pattern.*

If every kite is embellished with either a chocolate- or vanilla-colored circular sector at its central vertex as in Figure 17(b), the 12 triangles of quarks become a fanciful toy model of the Quantum ChromoDynamic *flavored* and *colored* structure [4, 6, 14] of the twelve nucleons of the ^{12}C nucleus.   Each kite represents a quark. According to QCD, a proton is composed of two *up* quarks, each with electric charge $+2/3$, and one *down* quark, each with electric charge $-1/3$, yielding a net electric charge of $+1$. A neutron, on the other hand, is composed of one *up* quark and two *down* quarks, yielding zero net charge. In quarks, electric charges (corresponding to flavor) are represented by the colors of the circular sectors: vanilla $= 2/3$ and chocolate $= -1/3$. In Figure 17(c), the electrical interaction energy between nucleons has been minimized

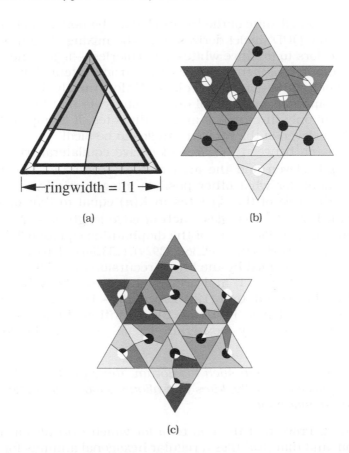

Figure 17. (a) Three kites and the associated trigonal ring. (b) Quarks (packaged). (See Color Plate IV(a).) (c) Quarks (pied). (See Color Plate IV(b).)

by placing a proton triangle in each of the six outer "petals" of the hexagram. In six of the twenty-eight pied solutions, the chocolate and vanilla sectors are grouped in patterns that satisfy the flavor requirements for the combinations of up and down quarks in protons and neutrons. Each of the other twenty-two pied solutions contains one or more fictitious nucleons with charge −1 or +2.

The pied coloring of every three-kite assembly is a metaphor for the QCD requirement that protons and neutrons be *color neutral*, i.e., that the net *color charge* associated with the strong interaction between quarks be zero. The packaged state of quarks represents a fanciful excited state of the ^{12}C nucleus. The notion of color neutrality in the QCD model derives from the mixing of light of three primary colors to produce white—i.e., colorless—light. The special unitary group SU(3) that defines the strong interaction actually calls for only three—not six—colors. Unfortunately, however, our fanciful computer modeling of quarks suggests that for every distribution of only three colors among the kites, it is impossible for more than eleven of the twelve triangles to be pied!

The match between the area of twelve equilateral triangles of edge length eleven and the area of the $\binom{9}{2}$ kites in $K(9)$ invites the question: for what other positive integer pair (k, n), besides $(1, 9)$, is the area of the $\binom{n}{2}$ kites in $K(n)$ equal to that of a hexagram tiled by $12k^2$ triangles, each of edge length $n + 2$? We find the following solutions (k, n) of the diophantine equation $36k^2 = \binom{n}{2}$: $(1, 9), (34, 289), (1155, 9801), (39236, 332929), (1332869, 11309769), \ldots$. The values of k are defined by the linear recursion $k_m = 34k_{m-1} - k_{m-2}$, $k_{-1} = 0$, $k_0 = 1$ ($m \geq 1$) and are generated by Chebyshev's polynomial of the second kind [12]. Since $n(n - 1) = 72k^2$, either $n \equiv 0 \pmod{3}$ or $n \equiv 1 \pmod{3}$. Hence the algorithm of Exercise 19 defines a tiling of every hexagram for $m \geq 1$ by the kites of $K(n)$.

Conjecture 6. *For every such hexagram, there exists a scheme for assigning six colors to the kites that allows a monochromatic-to-pied triangle transformation.*

Exercise 22. Prove that there is no n for which $K(n)$ tiles a regular hexagon, and that $K(n)$ tiles a regular hexagonal annulus for $n = 9$, 28, 37, 45, 64, 72, 73, 81, 100, $\ldots$.

If we restrict the size of the tiling arena to a reasonable value but reduce its symmetry from $d6$ to $d3$, we discover a three-color puzzle, *Cache-22*, that is composed of the sixty-six kites of $K(12)$. The object here is to transform the packaged pattern of Figure 18(a), in which twenty-one triangles are monochromatic and one triangle is pied, into a *half-pied* pattern like that of Figure 18(b), in which each of the twenty-one triangles contains two kites of one color and one kite of a second color.[1] It is unknown whether three colors can

---

[1]P. A. MacMahon, the combinatorics pioneer, investigated the assignment of n colors to the m congruent "compartments" of every triangle, hexagon, or square in a periodic tiling, where $m = 3$, 4, or 6, respectively [7]. The total number of possible color combinations was required to equal the total number of compartments.

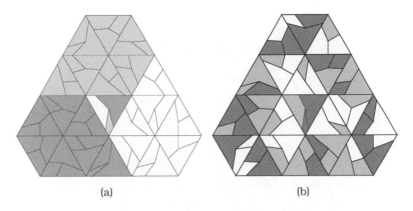

Figure 18. (a) Cache-22 (packaged). (b) Cache-22 (half-pied).

be distributed among the kites of $K(12)$ in a pattern that allows *perfectly half-pied* tilings, in which every color appears once in each of seven triangles and twice in each of seven other triangles. Also unexplored are tilings by the 45 kites of $K(10)$ of an equilateral triangular annulus with a central triangular hole.

Cache-51 is composed of the 153 kites of $K(18)$.

Conjecture 7. *Three colors can be assigned to the fifty-one monochromatic triangles of the packaged tiling of Cache-51 in a distribution that allows rearrangements in which every triangle is half-pied. The d3-symmetric tiling arena is a semi-regular hexagonal annulus with a triangular hole at its center (see Figure 19).*

Exercise 23. Let $n = 2(\lambda + 1)^2$ $(\lambda = 1, 2, ...)$. Prove that the area of the $d3$-symmetric semi-regular hexagon H_λ, which has alternating edge lengths $\alpha = \binom{\lambda}{2}$ and $\beta = \binom{\lambda+1}{2}$, is equal to the combined area of the $\binom{n}{2}$ kites of $K(n)$, and that H_λ can be tiled by $K(n)$.

For some values of n, he obtained a symmetrical tiling arena. In other research, MacMahon studied tilings by sets of non-congruent regular polygons. Here we focus on sets of non-congruent quadrangular kites. We first identify values of n for which the combined area of the $\binom{n}{2}$ kites in $K(n)$ is the same as that of some symmetrical arena, and then choose for the number of colors assigned to the $\frac{1}{3}\binom{n}{2}$ monochromatic triangles, each tiled by three kites, a value that defines a puzzle that is both challenging and esthetically satisfying. Deciding which kites receive which colors depends on whether the monochromatic triangles are to be rearranged into (a) pied or (b) half-pied triangles. In every case, it is quite easy to reassemble the monochromatic triangles, but the inverse transformations (monochromatic $\Rightarrow$ pied and monochromatic $\Rightarrow$ half-pied) are challenging.

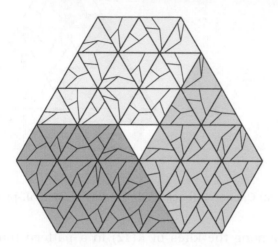

Figure 19. Cache-51 (packaged).

Exercise 24. Prove that the $\binom{n}{2}$ kites of $K(n)$ tile an equilateral triangular arena (a) for $n = 3 \times 1^2$, 3×9^2, 3×89^2, 3×881^2, 3×8721^2, ..., for which $n \equiv 0 \pmod 3$, and (b) for $n = 5^2$, 49^2, 485^2, 4801^2, 47525^2, ..., for which $n \equiv 1 \pmod 3$ [13]. How would you arrange the 100 monochromatic triangles of $K(25)$ colored with five colors?

Incubus and Cáscara

The volume of the lomino $[i, j]$ is equal to the *sum* $i + j + 1$. We now describe two cube puzzles—Incubus and Cáscara—in which the volume of each piece is equal to the *product*, not the sum, of integer indices i and j. In the case of lominoes, i_{min} and j_{min} were each set equal to two so that every piece would be L-shaped. Allowing indices equal to one would have meant introducing straight bar-shaped pieces, thereby making tiling and packing problems much too easy. For cube puzzles like Incubus or Cáscara, if i_{min} and j_{min} were equal to one, some pieces would be so small that solving the puzzles would become trivial. Every index pair for Incubus is an entry in $T(5)$ (see Figure 20). For Cáscara, every index pair is an entry in $S(7)$ (see Figure 22). We describe Incubus first.

$$(2,2)\ (2,3)\ (2,4)\ (2,5)$$
$$(3,3)\ (3,4)\ (3,5)$$
$$(4,4)\ (4,5)$$
$$(5,5)$$

Figure 20. $T(5)$.

Figure 21. $B_T(5)$: the ten blocks of Incubus, defined by $T(5)$.

$$(2,2)\ (2,3)\ (2,4)\ (2,5)\ (2,6)\ (2,7)$$
$$(3,2)\ (3,3)\ (3,4)\ (3,5)\ (3,6)\ (3,7)$$
$$(4,2)\ (4,3)\ (4,4)\ (4,5)\ (4,6)\ (4,7)$$
$$(5,2)\ (5,3)\ (5,4)\ (5,5)\ (5,6)\ (5,7)$$
$$(6,2)\ (6,3)\ (6,4)\ (6,5)\ (6,6)\ (6,7)$$
$$(7,2)\ (7,3)\ (7,4)\ (7,5)\ (7,6)\ (7,7)$$

Figure 22. The square array $S(7)$.

For some $n \geq 2$, construct a $1 \times i \times j$ rectangular block $[ij]$ for every entry (i,j) in $T(n)$. These $\binom{n}{2}$ blocks, each of volume ij, have total volume $V_{\text{blocks}}(n) = (n-1)n(n+1)(3n+10)/24$.

It is natural to wonder whether there exist values of n for which $V_{blocks}(n)$ is equal to the cube of some positive integer, i.e., whether there are solutions in positive integers (n,m) of the following diophantine equation:

$$(n-1)n(n+1)(3n+10) = 24m^3. \qquad (4)$$

After noticing the solution $(n,m) = (5,5)$, I extended the search to values of n in the many thousands and found no other solutions. It happens that the set $B_T(5)$ of ten blocks corresponding to $T(5)$ (see Figures 20–21) does admit a packing of the $5 \times 5 \times 5$ cube, but I was intrigued to find that many people (including me) were not able to discover it without a struggle. If you don't count trivial rearrangements of convex subsets of pieces, the packing can be described as unique.

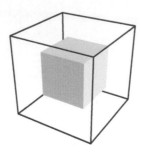

Figure 23. A $5 \times 5 \times 5$ cube embedded in $9 \times 9 \times 9$ cube.

Exercise 25. Pack a $5 \times 5 \times 5$ cube with the ten blocks of Incubus.

I emailed Benne de Weger, an expert in the design of algorithms for solving diophantine equations [15], asking him to investigate both Equation (4) and also several related diophantine equations, including Equation (5). He generously obliged!

$$(n-1)^2(n+2)^2 = 4m^3. \tag{5}$$

The left side of Equation (5) is the volume of the set $B_S(n)$ of blocks defined by $S(n)$, shown in Figure 22 for $n = 7$. After I told Benne about the solutions $(n, m) = (5, 5)$ for Equation (4) and $(n, m) = (7, 9)$ and $(n, m) = (126, 400)$ for Equation (5), he proved that neither equation has any other solutions [16]. The solution $(n, m) = (126, 400)$ for Equation (5), which (with apologies to Tartini) I have dubbed La Folia, defines the set $B_S(126)$ composed of 15,625 ($= 5^6$) blocks. (If one ignores the preposterous logistical problems of packing an actual $400 \times 400 \times 400$ cube with these blocks and considers only virtual packings, it is found that there is a simple solution, which is described below.)

The 36-block set $B_S(7)$ defined by $S(7)$ is dubbed Cáscara (Spanish for nutshell), because it suggests the following challenge: pack a $9 \times 9 \times 9$ cube that contains a concentric cube of edge length five or seven (see Figures 23–24). The packing for the outer shell that encloses a $5 \times 5 \times 5$ cube (see Figure 24) is one of several solutions discovered by Dave Aubertin [1] in a backtracking search. (Neither he nor I had found a packing by hand.) The 26 blocks in this shell are the pieces of $B_S(7)$ that do not belong to $B_T(5)$. Dave also found solutions for the version of Cáscara that contains a centered

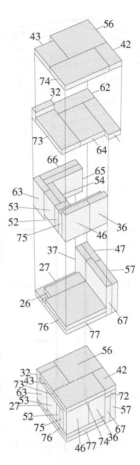

Figure 24. Outer shell of a $9 \times 9 \times 9$ cube that contains a centered $5 \times 5 \times 5$ cube.

$7 \times 7 \times 7$ cube. Figures 25–26 show a stratified packing of a $9 \times 9 \times 9$ cube by the blocks of Cáscara, distributed four-to-a-square among nine 9×9 stacked squares.

A *greedy algorithm*[2] is effective for packing a stratified cube composed of either the nine 9×9 squares of Cáscara or the four

---

[2]A greedy algorithm may be roughly characterized as a procedure in which the largest possible piece of the problem is solved at each step.

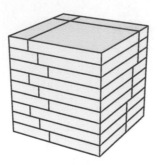

Figure 25. Stratified $9 \times 9 \times 9$ cube.

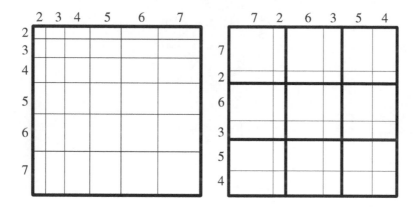

Figure 26. Outlines of the 36 blocks of $B_S(7)$ (Cáscara). The left pattern matches the entries in $S(7)$. The right pattern is a tiling of nine squares.

hundred 400×400 squares of La Folia. There is a square array for La Folia that is analogous to the one at the right in Figure 26, but the number of blocks per square is not constant. It increases from $4^2 = 16$ in the upper left square to $23^2 = 529$ in the lower right square.

James Dalgety calls a logically complete set of puzzle pieces an "English Selection" [2]. Although Incubus and Cáscara would be logically *more* complete if they included blocks of unit width, they would then be too easy. Children of age five or so, however, find the following $4 \times 4 \times 4$ cube puzzle, composed of nine rectangular blocks—including three of unit width—quite challenging: 1×2, 1×3, 1×4, 2×2, 2×3, 2×4, 3×3, 3×4, and 4×4. Dave

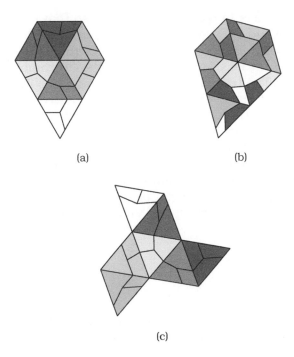

(a)   (b)

(c)

Figure 27. (a) Prime (packaged). (b) Prime (pied). (c) *Shuriken*.

Aubertin used his backtracking program to identify the 900 or so solutions [1]. Only rarely have I seen a child place the largest pieces first when attempting to pack this cube. Most little children immediately place the smallest pieces first, perhaps because they're confident *they will fit!*

As a tribute to Martin Gardner, puzzler and anagrammatist extraordinaire, let us now consider a small seven-triangle memento called Prime. Its level of difficulty—compared to Cache-22, for example—is suggested by an anagram of "Seven Triangles," its alternative name:

SEVEN TRIANGLES ⇔ LESS ENERVATING.

Each of the seven differently colored monochromatic triangles in the packaged tiling of Prime contains three of the twenty-one kites of $K(7)$. Besides tiling the outline of the *prime* symbol ('), the pieces also tile the outline of a $c3$-symmetric *shuriken* (ninja dart).

The object of Prime is to rearrange the kites so that every triangle is pied. Among the 27 triangles of the three possible solutions, nine of the $\binom{7}{3}= 35$ different combinations of three colors chosen from seven occur at least once, and 14 of the 35 combinations of three kites that tile a triangle of edge length nine (see Equation (2)) also occur at least once. For larger sets like CACHE-22, printing the edge lengths of the two outer edges on each kite makes it is easier to distinguish one kite from another, but the kites of Prime are sufficiently dissimilar to be identified easily at a glance.

There are seven other ways to tile seven monochromatic triangles by the 21 kites of $K(7)$ besides that of Figure 27, but none allows a rearrangement into seven pied triangles whose color distribution defines the Steiner triple system S(7) [10].

Solutions

Exercise 4

The signatures of the two dual corrals of Figure 28 are $\langle 7,2][3,2]$ $[5,6][7,8][7,6][3,8][3,4][5,4\rangle$ and $\langle 3,8][7,8][5,4][3,2][3,4][7,2][7,6][5,6\rangle$.

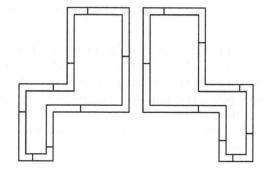

Figure 28. Oppositely congruent dual corrals.

Exercise 7

Hint for Exercise 7: The following algorithm is found to be effective in the search for a matched fence tiled by the $N(n) = \binom{n}{2}$ lominoes of Ln, for $n = 8k$ $(k = 1, 2, \ldots)$.

1. Construct a *matched signature* $\langle [a_1, a_2][a_2, a_3] \cdots [a_{N-1}, a_N],$ $[a_N, a_1] \rangle$.

2. Apply a CCW sequence of L (left) and R (right) turns, with $\Sigma L = \Sigma R + 4$, to the $N(n)$ consecutive lominoes in the signature of Step 1, thereby defining the orientation of the two arms of every lomino in the sequence.

3. If the turn sequence in Step 2 fails to produce a closed fence, replace one or more subsequences of turns (e.g., LR or LRRL) with *reversed* subsequences (RL or RLLR).

4. If repeated executions of Step 3 fail to yield a closed fence, return to Step 1.

Experience demonstrates that Step 1 may be implemented as follows:

(a) Select a lomino of Ln at random as the first piece in the signature.

(b) Select the next piece at random from the matching pieces not yet selected.

(c) Repeat Step (b) until either

  (i) all $N(n)$ pieces have been selected, and a cyclic signature is obtained, or

  (ii) there are no matching pieces among the pieces not yet selected, and the selection halts. Return to (a).

A simple way to implement Step 1 is suggested by the orderly structure of $T(n)$ and is illustrated in Figure 30 for L8. We denote the resulting signature as a *standard matched signature*. The matched fence in Figure 29 has a standard matched signature.

Exercise 8

A proof by Greg Martin [8] for standard sawtooths is contained in [11], which is a 150-page color-illustrated CD manual for a canonically colored set of the 32 pieces of L8$^+$. The set and manual can be purchased by writing to me at 316 W. Oak St., Carbondale, IL 62901.

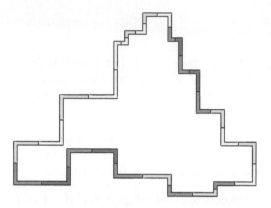

Figure 29. Matched fence of order 8.

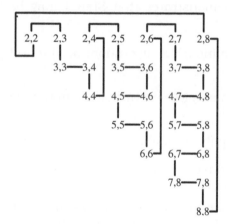

Figure 30. Standard matched signature for $n = 8$.

Exercise 17

Let us denote the square rings of ringwidth five, six, and seven as R5, R6, and R7, respectively. Because there are no pieces in L5 with armlength six or seven, it is impossible for either R6 or R7 to be a 2-ring or 3-ring. Hence both are 4-rings. That leaves only two pieces for R5, which is therefore a 2-ring, with signature $\langle [5,5][0,1][4,4][1,0] \rangle$. Since none of the pieces $[2,5]$, $[3,5]$, or $[4,5]$ can fit in a 4-ring of ringwidth six, they must belong to R7. Since the combined volume of these three pieces is twenty-one and the volume of R7 is twenty-four, the fourth piece in R7 must be [2,2], the only piece with volume three. The signature of R7 is therefore either $\langle [5,4][3,5][2,2][5,2] \rangle$ or $\langle [5,4][3,5][2,5][2,2] \rangle$. The four remaining pieces define R6 as $\langle [4,3][3,3][3,2][4,2] \rangle$.

Bibliography

[1] David Aubertin. Personal communication, 1990.

[2] Nick Baxter. "The Burnside Di-Lemma: Combinatorics and Puzzle Symmetry." In *Tribute to a Mathemagician*, edited by Barry Cipra et al., pp. 199–210. Wellesley, MA: A K Peters, Ltd., 2005.

[3] Martin Gardner. *Knotted Doughnuts and Other Mathematical Entertainments.* New York: W. H. Freeman and Co., 1986.

[4] Murray Gell-Mann. *The Quark and the Jaguar.* New York: Henry Holt and Co., 1994.

[5] Solomon W. Golomb. *Polyominoes: Puzzles, Patterns, Problems, and Packings*, Revised and Expanded edition. Princeton, NJ: Princeton University Press, 1994.

[6] Brian Greenc. *The Elegant Universe.* New Tork: Vintage Books, 2000.

[7] Percy Alexander MacMahon. *New Mathematical Pastimes* (Introduction by Paul Garcia), Reprint of the 1930 Edition. St Albans, UK: Tarquin Reprints, 2004.

[8] Greg Martin. Personal communication, 1999.

[9] Ed Pegg Jr. Personal communication, 2006.

[10] W. W. Rouse Ball and H. S. M.Coxeter. *Mathematical Recreations and Essays*, Thirteenth Edition. New York: Dover Publications, 1987.

[11] Alan H. Schoen. "Lominoes." Available at http://www.schoenpuzzles.com, to appear.

[12] Neil J. A. Sloane. "Sequence A029547," *On-Line Encyclopedia of Integer Sequences*. Available at http://www.research.att.com/~njas/sequences/index.html, 2008.

[13] Neil J. A. Sloane. "Sequences A072256 and A001079." *On-Line Encyclopedia of Integer Sequences*. Available at http://www.research.att.com/~njas/sequences/index.html, 2008.

[14] Stanford Linear Accelerator Center. "Quarks" (color charge theory). Available at http://www2.slac.stanford.edu/vvc/theory/quarks.html, 2007.

[15] B.M.M. de Weger. *Algorithms for Diophantine Equations*, CWI Tract 65. Amsterdam: Stichting Mathematisch Centrum Centrum voor Wiskunde en Informatica, 1989.

[16] B.M.M. de Weger. Personal communication, 1999.

The Hexa-Dodeca-Flexagon

Ann Schwartz and Jeff Rutzky

What is it? The *hexa-dodeca-flexagon*, or 12-gon, is a flexagon composed of 30°/60°/90° triangles. When it is first folded up, it looks like a hexagon divided neatly into 12 pie slices.

The 12-gon follows the basic flexagon rules. In many respects the 12-gon is deliciously similar to its cousins, the hexaflexagon and the 8-gon, the flexagon made from isosceles right triangles.[1] Like them, the hexa-dodeca-flexagon follows these general rules:

- It can be made from a straight strip of paper.

- The number of triangles needed on each side of the strip follows the formula used for the hexaflexagon and 8-gon:

$$\frac{(\text{number of triangles per face})(\text{number of faces})}{2} + 1$$

- The layers of thicknesses in adjacent pie sections of the hexagon, also called "pats," add up to the total number of faces in the flexagon—although as we'll see later, the 12-gon flexes

---

[1] For a full discussion of the 8-gon, see [1].

into states that are not radial, and in which the term "pie section" does not apply. Still, the layers of triangles in adjacent pats—with one notable exception—will always add up to the total number of faces.

- Like the hexa-hexa-flexagon, the 12-gon has different classes of faces; some faces appear more often during flexing.

- Like the 8-gon, the dodeca can be manipulated with a "pass-through" flex, during which the flexagon seems to do a little somersault before settling into a new face. And also like the 8-gon, the 12-gon can be manipulated with a "reverse pass-through" flex, during which pairs of pats can be lifted up to progress to a new face.[2]

The 12-gon has some new tricks of its own. Among its innovations:

- It has hidden triangles on its template that very rarely appear after the flexagon is folded up.

- It combines triangles from different faces to form hybrid faces.

- It flexes into different flat shapes.

- It can be flexed along at least four different combinations of creases.

- It has a special set of triangles, called "Toggle Triangles," that change how the hybrid faces are formed.

- Rather than having just two different classes of faces like the hexa-hexaflexagon, it has four.

Templates and Folding

In the template in Figure 1, notice that there are a total of 37 triangles; six colors, one for each basic face (indicated by the letters R for red, O for orange, Y for yellow, G for Green, B for blue, and V for violet); and groups of unmarked triangles arranged in kite-like patterns on both sides of the strip. These triangles, for the most

---

[2]For a discussion of the pass-through and reverse pass-through flexes, see [1].

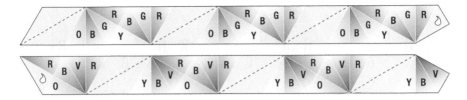

Figure 1. Template with 37 triangles.

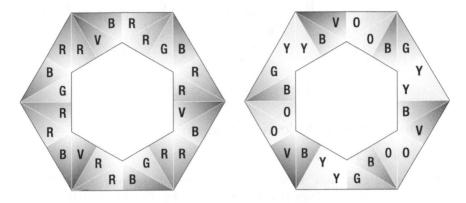

Figure 2. Hollow hexagon.

part, stay hidden. Because of them, some of the faces have only six triangles and must "borrow" triangles from other colors to form a complete surface.

Fold the unmarked triangles lengthwise along the dotted lines on the template, and place the triangles with the droplets face-to-face. The resulting shape is a hollow hexagon, as shown in Figure 2.

Notice that the orange and yellow triangles now appear in adjacent pairs. Fold each orange and each yellow triangle onto its twin. You'll now have a row of zigzags, as shown in Figure 3.

Now fold each violet and each green triangle onto *its* adjacent twin. You'll be forming a hexagon with 12 red triangles on one side and 12 blue triangles on the other side. Glue the two triangles with the droplet face to face.

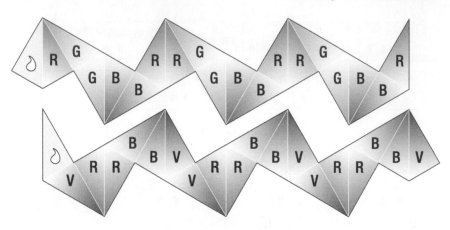

Figure 3. Row of zigzags.

Flexing the Hexa-Dodeca-Flexagon

Corner Flexing along Three Axes

Flex the 12-gon just as you would a hexaflexagon, using three equidistant creases that run from the center to the corners of the figure. You'll produce four different positions. The numbers in parentheses on the diagrams in Figures 4–7 indicate the number of layers in each pat.

1. *Radial–1.* The triangles are arranged radially around the center of the hexagon, with alternating thick and thin pats. Both sets of axes can always be flexed to a new face. The Radial–1 position is illustrated in Figure 4.

2. *Radial–2.* Like in the Radial–1, the triangles are arranged radially, but here the thick pats appear side by side and are followed by two adjacent thin pats. Radial–2 will only flex along one axis. As with Radial–1, Radial–2 forms four unique surfaces. The creases to flex along always lie between like-colored triangles. The Radial–2 position is shown in Figure 5.

3. *Alternate Hexagon.* It's a hexagon, all right, but its triangles are arranged quite differently, as you can see in Figure 6. The Alternate Hexagon always flexes along both sets of axes. It forms seven unique faces, including one pair that is pattern-inversed.

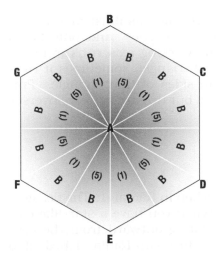

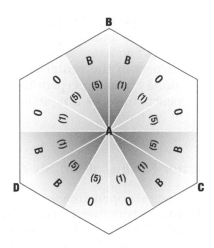

Figure 4. The 12-gon in Radial–1 position. Flex along **AB**, **AD**, and **AF**, or **AC**, **AE**, and **AG**. In the illustrated model, the Radial–1 positions have faces that are either all blue, all red, orange and yellow, or green and violet.

Figure 5. The 12-gon in Radial–2 position. Flex along **AB**, **AC**, and **AD** only. In the illustrated model, Radial–2 always combines triangles from two different basic faces in the pattern shown. (See Color Plate X(a).)

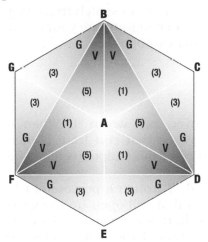

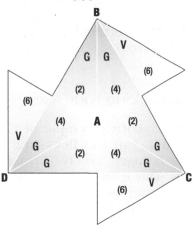

Figure 6. The 12-gon in the Alternate Hexagon position. Flex along **AB**, **AD**, and **AF**; or **AC**, **AE**, and **AG**. (See Color Plate X(b).)

Figure 7. The 12-gon in Propeller Position. Flex along **AB**, **AC**, and **AD** only. (See Color Plate X(c).)

4. *The Propeller.* The fascinating Propeller, as illustrated in Figure 7, is made up of only nine triangles, and only flexes along one axis. Sometimes the working axis runs from the corners of the center equilateral triangle; sometimes it runs along creases perpendicular to sides of that triangle. Propellers have 14 unique faces. All but two are pattern inverses of each other.

Corner Flexing along Six Axes

Flex along six open creases at once when in Radial–1. The flexagon will start to turn itself inside out. When you have a star-like pattern of six horizontal triangles radiating outward from the center, stop flexing. For instance, if you start with Radial–1 Red, stop when you see violet and green triangles. Hold the flexagon so that a pair of either violet or green triangles, horizontal and with the hypotenuses pointing up, is facing you. The triangles should have an open crease along the tops of the hypotenuses. Open it and pull a pair of triangles down. You should now see four blue triangles. Work your way around the star, pulling down the green and violet triangles and exposing blue ones. Then flex through the center. You'll find that this is a way—a difficult way—to switch from "Regular" into "Wormhole" mode, but you can congratulate yourself for having executed a difficult reverse pass-through flex.

Flexing along Four Axes

This is another complicated maneuver, but worth it. When in Radial–1, you can flex along two creases that run from corners of the hexagon, and two creases that run from midpoints along the sides. This maneuver is depicted in Figure 8.

For example, start in Radial–1 Red. Through a process of trial and error, find the set of four creases that will open up to expose three green and three violet triangles. In this state, you will see two red triangles protruding at opposite ends of the figure. Push all the triangles down until the short legs of these red triangles meet at the bottom and center of the flexagon, and the green and violet triangles are perpendicular to the ground with their hypotenuses pointing up. The next maneuver is similar to one described for

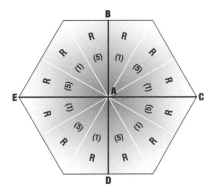

Figure 8. Flexing along the four axes. Pinch back along **AB**, **AC**, **AD**, and **AE**.

flexing along six axes. The triangles should have an open crease along the hypotenuses. Open a crease and pull two violet triangles down until they are face to face against the red one. You should now see four blue triangles. Work your way around the star, pulling down pairs of violet and green triangles and covering them with blue ones. At this point, you'll have a star-like three-dimensional figure with four blue rays, with the long legs of the blue triangles on top. Now flex through the center of the flexagon.

The result is an altogether new position, Radial–3, which is shown in Figure 9. Unlike the positions seen so far, it combines *three* colors in one surface. There are at least two pairs of Radial–3 faces.

Note how the pat structure of Radial–3 is different from those of Radial–1 and –2, and that it is impossible to flex along three creases from Radial–3.

Other things can happen when you're flexing along four axes. An additional pair of triangles can slip out, throwing you into Flex– 6. And from one point in the flex–4³ procedure, the 12-gon can open up into a flat rhombus composed of 12 red triangles.

You can also flex the 12-gon into a stable zigzag shape. Start in Radial-1 using the face that has only blue triangles. Pick four equi- distant creases that *don't* open in the center. Two of the creases are

---

[3]I know it's possible to go from Radial–1 in Regular Mode to the first pair of Flex– 4 faces, and then flex to Radial–1 in Wormhole Mode, and then to the other pair of Flex–4 faces. I also know that you can use the Toggle Triangles to go directly from one Flex–4 face to another one. Just don't ask me yet how it's done!

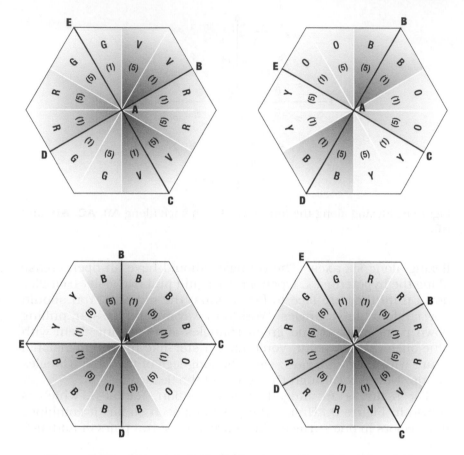

Figure 9. The 12-gon in Radial–3 position. (See Color Plate X(d).)

blocked, but two will open! Pull down the blue triangles along the two creases. You'll now have two large diamonds that have green and violet triangles on one side and blue triangles on the other. Pull the diamonds away from each other to flatten the mountain fold between them into a blue equilateral triangle. This new triangle has an open crease at its top that can be pulled down. When it is, you'll have three diamonds in a zigzag shape that are either blue or violet and green. To get back to Radial–1, fold down the green and violet triangles so that the like colors are face to face. I wonder... can this operation be considered flexing along *two axes*?

Midpoint Flexing

From certain positions you can flex along equidistant creases that run from the midpoints of the sides of the flexagon to the center. Go to the Radial–1 position. On the side that has only blue triangles, flex along three midpoint creases. The flexagon will open up and show either six green center triangles surrounded by six blue outer triangles, or six violet triangles surrounded by blue triangles. The flexagon will not be able to flex flat at this point, so keep flexing until the green or violet triangles are horizontal with their hypotenuses facing down and the blue triangles are below them. Open up the center of the flexagon. If you do this right, you will have produced a red equilateral triangle. Flip it over and you should see another triangle composed of alternating blue and green or blue and violet triangles. Note how this oddball position breaks two previous rules: it combines colors (faces) that have never combined before, and it exhibits a pat structure in which adjacent thicknesses add up to 12, *twice* the number of faces. With either the blue and green equilateral triangle, or the blue and violet one facing up, flex along the triangle's long axes using a reverse pass-through flex. If you do this correctly, you will have flexed to either an all-yellow equilateral triangle or an all-orange one. I have found at least ten differently colored equilateral triangles, including one made up of alternating blue and red 30°/60°/90° triangles. Flexing along the short axes produces cups and pyramids. To find all the triangles you can, also flex along the midpoints of the faces in Radial-2.

Another Shape and a Surprise

The hexa-dodeca-flexagon has at least one more trick up its sleeve. Starting in a Radial–1 position, locate four creases that run from the midpoints of four sides of the hexagon and are separated by two 60° angles and two 120° angles. Fold back along these creases, and while the flexagon is opening, pull along two opposite triangles and twist slightly. This takes some practice. Your goal is to produce a parallelogram composed of 12 triangles. This parallelogram can be flexed! Fold back along the long diagonal. The flexagon will now form a kind of cup. With just a slight bit of bending, force the triangles at the bottom of the cup up and through the center of the structure. The triangles that formed the bottom of the cup will now be pointed toward you. Separate the two halves of the new parallelogram the way you would open a book. You'll have produced a parallelogram with a new face, probably one with five of the flexagon's basic six faces on one surface. Repeat—flex the parallelogram again. The next cup reveals—finally—the triangles that normally stay hidden. You can uncover all the hidden triangles with this challenging maneuver.

Regular and Wormhole Modes and Toggle Triangles

When you're corner flexing along three axes, the 12-gon will exhibit different color combinations and color arrangements depending on how certain triangles are positioned. When the model is in the Alternate Hexagon position, and all the triangles are blue,[4] there are three pairs of triangles along the outside of the hexagon that are unattached at one corner. These are the Toggle Triangles. Leaving them in the position that maintains an all-blue Alternate Hexagon keeps you in the Regular Mode. To enter the Wormhole and get some new faces, fold back each pair of Toggle Triangles onto the blue triangles adjacent to them. You should now see a green and violet Propeller. Flex the Propeller and note the new color combinations. Most noticeably, you'll find faces that mix orange and yellow, and green and violet for the first time.

Different Classes of Faces

As shown earlier, the colors red and blue have 12 triangles apiece and green, violet, orange, and yellow have only six triangles each. But there are further distinctions. All six violet triangles appear on one side of the strip, as do the green ones. Orange and yellow, however, each have three triangles on both sides of the template. These differences affect the behavior of these colors in the flexagon. In addition, where the red and blue triangles appear in relation to the "lesser" colors on the template affects red and blue as well. The six basic faces break up into four classes:

---

[4]Toggle Triangles are also found when the Propeller in the model has blue triangles only. When in Wormhole Mode, the Toggle Triangles are violet and green.

1. *Alpha face.* This is blue in the model. Alpha is one of the two faces that has 12 triangles. It is the only one that can exclusively form the Radial–1, Alternate Hexagon, and Propeller positions. Toggle Triangles are always present when Alpha is in the Alternate Hexagon and the Propeller. Alpha only combines with Delta faces when flexing along three corner creases or a combination of two corner and two midpoint creases.

2. *Beta face.* This is red in the model. It is one of the two faces that has 12 triangles. It can exclusively form Radial–1 and the Propeller. Beta only combines with Gamma faces when flexing along three corner creases or a combination of two corner and two midpoint creases.

3. *Gamma faces.* These are violet and green in the model. Having only six triangles each, they must borrow triangles in order to form a position with 12 triangles, or the Propeller, which has nine triangles. They can combine with each other and Beta when flexing along three corner creases or a combination of two corner and two midpoint creases. In Wormhole Mode, both colors will be on the Toggle Triangles.

4. *Delta faces.* These are orange and yellow in the model. Like Gamma faces, they must combine with other colors. They mix only with Alpha or each other when flexing along three corner creases or a combination of two corner and two midpoint creases.

Tuckerman Traverse

Named after Bryant Tuckerman who discovered it, this series of flexes reveals every face of the hexa-hexa-flexagon with the least number of flexes. Repeating cycles of the hexa-dodeca-flexagon are also possible. Twelve flexes in Regular Mode are necessary to reveal every color at least one time. It is worth noting that Delta colors appear only once, while Gamma colors appear at least twice. Twelve flexes also produce a repeating cycle in Wormhole Mode.

For More Information

Email Ann Schwartz at annschwartz101@yahoo.com, and feel free to join the Yahoo! Flexagon Lovers Users group at http://groups.yahoo.com/group/Flexagon_Lovers.

Bibliography

[1] Peter Hilton and Jean Pedersen. *Build Your Own Polyhedra*, Second Edition. Reading, MA: Addison Wesley, 1994.

Golomb, Gardner, Benjamin and Jones: Midwives to a Puzzle

Norton Starr

While giving a classroom lecture on mathematical induction, I recalled once having heard a surprising application of induction to a tiling problem. Supplementing my prepared examples, I ad-libbed the argument, a classic one due to Solomon Golomb. This note is about my puzzle demonstrating his technique for tiling checkerboards with *trominoes*, a type of polyomino.

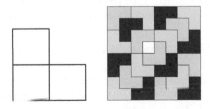

Figure 1. A tromino (left) and a tiling of an 8 × 8 board (almost) with trominoes (right).

269

Golomb's founding and naming the field of polyominoes are widely known [9, 11]. So is Martin Gardner's *Scientific American* midwifery to that area of combinatorial geometry [6, 7]. Less familiar is Gardner's vital role in Joseph Madachy's launch of the *Recreational Mathematics Magazine*, the first popular outlet for Golomb's articles on polyominoes [10]. Gardner's assistance is described below, along with a sketch of some of the subsequent history of Golomb's elegant tromino tiling work. That result took on a life of its own and continues to be discussed and generalized in books and articles, as we will show. First I note how it led to a popular puzzle, and at the end I propose some extensions to that puzzle.

At the January 2000 annual meeting of the Mathematical Association of America, Arthur Benjamin received the Haimo award for distinguished college teaching. In his acceptance speech he sketched his favorite proof by induction, the argument due to Golomb. Its reasoning assures the possibility of using L-shaped tiles ("trominoes") to fill ("tile") a $2^n \times 2^n$ celled square, one of whose cells is already occupied (a "deficient" square). Three years later this was the memorable application that I presented in a lecture and that stimulated my development of a tromino puzzle to demonstrate mathematical induction. Wanting to show my students a physical example of the process, I contacted the most likely source, Kadon Enterprises, the leading developer and maker of polyform tiling puzzles and games. Kadon did not have it among their offerings, so I commissioned them to make a few of size 8×8 for me. Kate Jones, their president, asked if they could add it to their product line. I readily agreed to this way of bringing the tromino puzzle to a broad audience. She improved my design by suggesting the use of more colors, by enhancing the brochure with strategy games and variations on the basic puzzle, and by proposing that users might also try to achieve tilings that are properly color-separated. These improvements made the puzzle more interesting than I had originally envisioned. Indeed, it has become one of Kadon's most popular products.

The proof of the tromino tiling was published by Golomb in his 1954 *American Mathematical Monthly* paper [9, p. 677], and provides a first-rate example of mathematical induction. Beyond the sheer elegance of the argument, it's an unusual instance of a nonnumerical application of the method. This stands in contrast to the examples and exercises typically found in textbook treatments of induction, which consist of a variety of formulas for finite sums, inequalities, and the like. (Mathematicians will appreciate Golomb's remark after his proof: "This proof is similar to the Bolzano-Weierstrass Theorem in two dimensions. We keep subdividing into quadrants until we locate a monomino, our discrete analogue of a cluster point." The Bolzano-Weierstrass Theorem is a cornerstone of mathematical analysis, underlying powerful theorems used in mathematics, economics, etc.)

The tromino tiling result Golomb proved in the *Monthly* was alluded to in Gardner's seminal *Scientific American* article of May, 1957 introducing polyominoes to the wider public [6]. For his first book of collected *Mathematical Games* columns, Gardner combined his checkerboard puzzle remarks in that May, 1957 column with the more complete treatment of pentominoes in the December, 1957 column [7] to form Chapter 13, "Polyominoes" [8]. His comment about tromino tiling in that chapter, not present in the two source columns, was: "an ingenious induction argument demonstrates that 21 right trominoes and one monomino will cover the 8-by-8 board regardless of where the monomino is placed" [8, p. 126]. The actual proof's first appearance in a popular medium was in Joseph Madachy's *Recreational Mathematics Magazine* (*RMM*), where Golomb included it in "The General Theory of Polyominoes," the first of his four *RMM* articles on that topic [10, pp. 5, 7].

It's interesting to note that Martin Gardner played a valuable role in the founding of *RMM*. His generous assistance is described by Joseph Madachy as follows [15]:

> I hope things go well at the event honoring Martin Gardner. In June 1960 I wrote to him asking if there were any publications devoted to recreational mathematics. He said that there were a few, but none devoted *solely* to rec. math. I wrote back stating that I was thinking about starting such a project and could he give me some leads to get me started. He replied with a list of people who could produce suitable material and, gloriously!!, he also sent me a big box of his correspondence and said I could lift off the addresses and send promotional litera ture to them. By February 1961, seven months later, I was at the Idaho Falls Post Office sending out Issue 1 of *RMM* with about 1200 subscribers. By the end of 1961 it was up to about 4000 subscribers. It only lasted for 14

issues—but it was fun (and frustrating). Then, in mid-1966, Greenwood Press wanted me (at Martin Gardner's recommendation) as Editor of a new rec. math. Journal. Six months later, out came *JRM* (talk about history repeating itself).

(This new journal was the *Journal of Recreational Mathematics*, from which Madachy retired in December 2000 "with two editors taking my place.")

The method of proof illustrated by the tromino puzzle, known as the "Principle of Mathematical Induction," dates back at least to medieval Hebrew and Islamic mathematics of 700 years ago [13]. It's summarized in the glossary of Golomb's book [11, p. 173] and less terse accounts abound in the literature (see, for instance, [5].) This type of reasoning is like a process for figuring out whether an infinite string of dominoes, each standing on end, is arranged in such a way that all of them must fall down if the first topples over. Suppose that when any domino falls down, it hits the next domino, pushing it over. If that is true for every domino, and if the very first domino falls over, then they all fall down. The same idea can be applied to tromino tiling of deficient checkerboards (boards with one cell occupied) of dimensions $2^k \times 2^k$. For the initial case $k = 1$, a 2×2 board with one of the four cells occupied can be tiled with the obvious placement of a single tromino. (This corresponds to being able to push the first domino over.) Next, let k be some positive integer and *assume*, for the sake of argument, that any deficient checkerboard of side length 2^k (having dimensions $2^k \times 2^k$), can be tiled with trominoes. Using this assumption, we now show how to tile a deficient board of side length 2^{k+1}. (This corresponds to showing that if any domino falls over, it causes the next domino to fall down as well.) The Principle of Mathematical Induction then guarantees that every deficient checkerboard having side length a power of 2 can be tiled with trominoes.

Denoting the side length 2^{k+1} by $2n$ (so $n = 2^k$), consider the deficient board of dimensions $2n \times 2n$, as shown on the left in Figure 2. The single occupied cell must lie in one of the four $n \times n$ quadrants of the board—in the illustration it's in the upper left quadrant. The other three $n \times n$ quadrants have all their $2^k \times 2^k$ cells unoccupied. Take a single tromino and place it at the center of the board so that each of these other three empty quadrants contains exactly one of the three squares of that tromino, as shown

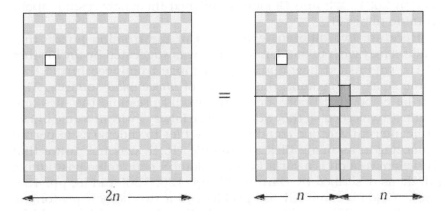

Figure 2. A deficient $2n \times 2n$ board. (Copyright © 2000 The Mathematical Association of America. From *Proof Without Words II* [18] with the permission of the Mathematical Association of America and Roger B. Nelsen.)

on the right in the figure. Thus each of the four $n \times n$ quadrants now is a deficient $2^k \times 2^k$ checkerboard. But we assumed any such deficient $2^k \times 2^k$ checkerboard can be tiled by trominoes! Thus *if* any deficient $2^k \times 2^k$ checkerboard can be tiled by trominoes, we've demonstrated how to tile any deficient $2^{k+1} \times 2^{k+1}$ checkerboard. This is Golomb's original argument. (Note that it provides an effective procedure for solving tromino puzzles, whether physical or online.)

Ross Honsberger gave perhaps the most lucid, detailed exposition of how Golomb's tiling technique can be applied to the 8×8 checkerboard, in his *Mathematical Gems II* [12, p. 61]. The most crystalline illustration of the general argument is Roger Nelsen's concise "proof without words," given in his second book of that title [18, p. 123], where Figure 2 appears.

Tromino tiling has inspired a stream of scholarship that continues to this day. Extensions of the basic puzzle are presented in the work of Chu and Johnsonbaugh [3, 4]. The topic is looked at from various angles in George Martin's book chronicling polyominoes generally [16]. Coloration problems for tromino tiling are treated by Mizniks, who acknowledges the Kadon Vee-21 color selection web page as inspiration for the research on which he reported in 2004 [17]. ("Vee-21" is Kadon's name for the 8×8 tromino puzzle.)

Also in 2004, Ash and Golomb published their new results about tromino tilings [1].

Here are two different extensions of the tromino puzzle that readers may wish to ponder. The first was suggested by my brother R. H., who asked how one might arrange trominoes in the 8×8 frame so as to maximize the number of *unoccupied* squares. This can be elaborated upon: one route would presume the tiles and frame are velcroed so they stay in place (always within frame lines), while alternatively one could allow the tiles to slide so as to permit squeezing in as many tiles as possible. R. H. was not aware that the velcro version is a variation on Golomb's pentomino positioning puzzle as described by Gardner [7, pp. 128, 130] and [8, pp. 133, 135]. Golomb extended the problem to a two-person pentomino game [7, p. 128] and [8, pp. 133–135], rules for which could be applied to the tromino puzzle as well. David Klarner reported on a two-person pentomino game, Pan Kāi (developed by Alex Randolph and issued in 1961 by Phillips Publishers), which included the following constraint, adaptable to the tromino context: "the most important rule is that it is forbidden to play a piece inside a closed region of the board if fewer than 5 cells would then remain unoccupied, unless the move exactly fills up the region." [14, p. 8]

Another direction is three dimensional. Consider a cube of side length 2^n, containing 2^{3n} unit cells, one of which is occupied (*single deficiency*.) Can the remaining cells be tiled with three dimensional trominoes (three cubes in an L-shape, with two of them meeting the third on two adjacent faces of the latter)? The necessary condition that $2^n = 3k + 1$ turns out to be sufficient as well. The case of a $4 \times 4 \times 4$ cube presents some modest challenges that may amuse young puzzlers.

A natural generalization of Golomb's result to cubes of side length 2^n fails, because the $8 \times 8 \times 8$ cube has $2 + 170 \times 3$ cells, so tromino tiling would require that two cells be individually occupied (*double deficiency*). However, with the aid of a strong induction, it can be shown that for each positive integer n, any singly deficient cube of side length 2^n may be tiled by trominoes if n is even, and any doubly deficient cube of side length 2^n may be tiled by trominoes if n is odd. When n is odd, the argument is complicated by the extra degree of freedom inherent in the arbitrary placement of a second single cube.

Far simpler is the following generalization, rather easily shown by a recursive method similar to that of Golomb: any cube of side length 4^n, one cell of which is occupied, can be tiled with trominoes.

The above three-dimensional results can be strengthened to the tromino tiling of cubes C of arbitrary side length m, provided they have the appropriate number of deficiencies:

- If m is a multiple of three, C can be tiled.

- If $m = 3k + 1$ and C is singly deficient, C can be tiled.

- If $m = 3k + 2$ and C is doubly deficient, C can be tiled.

The first claim is easy (see Exercise 6.10, pp. 59 and 64 of [2]), the second not too difficult, and the third rather complicated [19].

Bibliography

[1] J. M. Ash and S. W. Golomb. "Tiling Deficient Rectangles with Trominoes." *Math. Mag.* 77 (2004), 46–55. Also available at http://math.depaul.edu/~mash/TileRec3b.pdf.

[2] V. Boltyanski and A. Soifer. *Geometrical Etudes in Combinatorial Mathematics.* Colorado Springs: Center for Excellence in Mathematical Education, 1991. (Under revision by A. Soifer. To appear in 2009 with Springer.)

[3] I.-P. Chu and R. Johnsonbaugh. "Tiling Deficient Boards with Trominoes." *Math. Mag.* 59 (1986), 34–40.

[4] I.-P. Chu and R. Johnsonbaugh. "Tiling Boards with Trominoes." *J. Recreational Math.* 18 (1985–86), 188–193.

[5] M. J. Erikson and J. Flowers. *Principles of Mathematical Problem Solving.* Upper Saddle River, NJ: Prentice Hall, 1999.

[6] M. Gardner. "About the remarkable similarity between the Icosian Game and the Tower of Hanoi." *Scientific American* 196 (May 1957), 150–156.

[7] M. Gardner. "More about complex dominoes, plus the answers to last month's puzzles." *Scientific American* 197 (December 1957), 126–140.

[8] M. Gardner. *The Scientific American Book of Mathematical Puzzles & Diversions.* New York: Simon and Schuster, 1959.

[9] S. W. Golomb. "Checker Boards and Polyominoes." *Amer. Math. Monthly* 61 (1954), 675–682.

[10] S. W. Golomb. "The General Theory of Polyominoes. Part I–Dominoes, Pentominoes and Checkerboards." *Recreational Math. Mag.* 4 (August 1961), 3–12.

[11] S. W. Golomb. *Polyominoes*. New York: Scribner's, 1965.

[12] R. Honsberger. *Mathematical Gems II.* Washington, DC: Mathematical Association of America, 1976.

[13] V. J. Katz. "Combinatorics and Induction in Medieval Hebrew and Islamic Mathematics." In *Vita Mathematica*, edited by R. Calinger, pp. 99–106. Washington, DC: Mathematical Association of America, 1996.

[14] D. Klarner. *Box-Packing Puzzles.* Multilithed notes, University of Waterloo, Ontario, 1973–74. (Portions of this are summarized in [12, Chapter 8].)

[15] J. S. Madachy. Personal correspondence, January 25, 2006.

[16] G. E. Martin. *Polyominoes. A Guide to Puzzles and Problems in Tiling.* Washington, DC: Mathematical Association of America, 1991.

[17] I. Mizniks. "Computer Analysis of the 3 Color Problem for V-Shapes." *Acta Societatis Mathematicae Latviensis*, Abstracts of the 5th Latvian Mathematical Conference, April 6–7, 2004, Daugavpils, Latvia. Available at http://www.de.dau.lv/matematika/lmb5/tezes/Mizniks.pdf, 2004.

[18] R. B. Nelsen. *Proofs Without Words II: More Exercises in Visual Thinking.* Washington, DC: Mathematical Association of America, 2000.

[19] N. Starr. "Tromino Tiling Deficient Cubes of Any Side Length." Item 0806.0524 at http://arxiv.org, 2008.

Walk, Slide, and Jump

James W. Stephens

Some would say that the goal of the puzzle designer is to entertain, not to frustrate. I will confess that I violate this principle all the time, and I derive a certain fiendish satisfaction from developing extraordinarily difficult puzzles. One key advantage of designing puzzles with the aid of computer software is that it allows me to generate outrageously difficult puzzles with ease.

A genre of puzzles that can pack a surprise punch is based on sliding coins. After seeing a sliding coin puzzle invented by Bob Hearn at Gathering for Gardner 6, I played around with this concept myself and developed a type of puzzle I called the Meandering Monk Maze. These puzzles evoke the peaceful image of monks walking through a garden, but are so difficult and frustrating that the peaceful reverie is quickly shattered. After trying to solve these puzzles myself I decided that a planet of our size needed only a limited number of Meandering Monk puzzles, and thus I created only three.

The hardest Meandering Monk Maze is shown in Figure 1. (A somewhat nicer color version of this puzzle is available in applet form at http://www.puzzlebeast.com.) The rules are as follows:

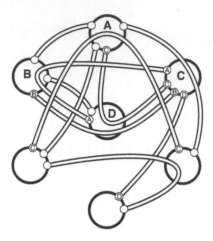

Figure 1. Meandering Monk Maze.

1. Use four coins or markers labeled A, B, C, and D to represent
   the four monks. Place the markers on the starting circles as
   labeled in Figure 1.

2. Each move consists of moving one monk (coin) along the complete length of one path connecting the large circles. A piece
   must always travel the complete length of a path and may not
   start a path in the middle or stop before completing the entire
   length of the path.

3. Paths that are labeled A, B, C, or D (as indicated by the small
   circles at the beginning and end of the path) may be traversed
   only by the monk with that label. Unlabeled paths are available for any monk to use.

4. A piece may not move along a path if any of the large circles
   crossed by the path are occupied by another monk.

5. The goal is to return the monks back to their starting spaces
   except with pieces and A and B swapped.

   The shortest solution for this puzzle is 174 moves.
   The concept of sliding things around has always been fertile
ground for innocent-but-deadly puzzles. Two deceptively difficult
sliding block puzzles are shown in Figure 2. The goal of each puzzle
is the same, return the pieces to their starting locations except with

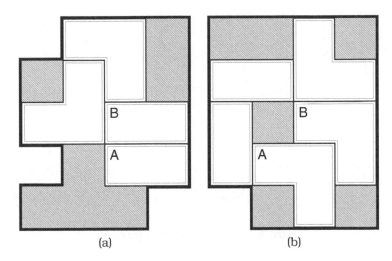

(a)   (b)

Figure 2. Two sliding block puzzles: (a) Picnic, (b) Fairly Complicated.

the positions of pieces A and B swapped. Rotation of pieces is not allowed.

Picnic requires 76 moves to solve, and Fairly Complicated takes a mind-numbing 148 moves. It is interesting to note that these puzzles share a trait with the Meandering Monk Mazes—the goal is to return all the pieces to the original configuration except for the interchange of two pieces. This attribute of the puzzles was not by intentional design but was a natural outcome of the computer generation process used to create the puzzles. It turned out that difficult puzzles found by the software all had this attribute.

In recent explorations I have identified a new format for puzzle-based torture, the jumping coin puzzle. I am delighted by the amount of sheer frustration that can be packed into a small grid with this type of puzzle. A number of relatively easy (e.g., less than 100 moves) jumping coin puzzles are available online at http://www.puzzlebeast.com under the name of the Fried Okra Perplexity. (A recipe for fried okra also is included on the website.)

Figure 3 shows two rather difficult jumping coin puzzles. The rules are as follows:

1. Use coins of two different sizes as pieces (e.g., quarters and pennies). One of the large coins needs to be identifiable as the target coin, perhaps by placing it heads up.

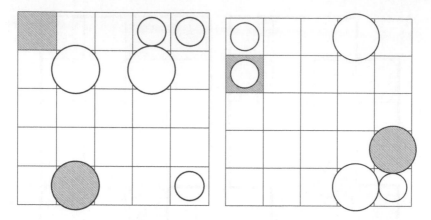

Figure 3. Two jumping coin puzzles.

2. Place the coins on the grid as illustrated. The shaded circle shows the starting location of the target coin.

3. Each move consists of selecting any coin on the grid and jumping it in an orthogonal direction over one *or more* adjacent coins landing on the next available empty space on the grid. Coins may not jump over vacant spaces. All coins are left on the grid after jumps.

4. The large coins are too big to fit next to one another orthogonally, and thus you may never have two large coins resting side by side in the grid. (However, large coins may rest diagonally adjacent to one another.)

5. The goal is to maneuver the target coin to the shaded target space on the grid.

You will note that only a limited number of spaces in the grid are reachable, which does not simplify the puzzles as much as you might think. The second puzzle is the hardest and requires 378 moves to solve. The first puzzle requires 168 moves to solve and is provided as a warm-up.

Polyomino Number Theory Developments

Robert Wainwright

Polyomino Number Theory is based on an idea presented by Solomon Golomb nearly 25 years ago [4]. Basically, the objective is to find a *least common multiple* of *compatible* polyomino pairs. At Gathering for Gardner 5, results of all pair-wise combinations of the smaller polyominoes (trominoes, tetrominoes, and pentominoes) were presented. These findings are described in several references [1–3]. A few interesting examples are shown in Figure 1. Figure 2 shows exchange puzzles from the April 2002 Gathering for Gardner and two gatherings since then. The instructions for these "Polyomino Number Theory Puzzle Challenges" are as follows:

Figure 1. Several interesting examples of compatible polyomino pairs: (a) $5X = 5N$, (b) $5W = 5T$, (c) $5Z = 5I$, and (d) $5L = 5X$.

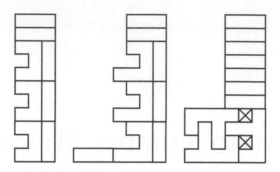

Figure 2. Exchange gifts from Gardner Gatherings 5, 6, and 7: (a) 5*U* and 3*I* (April 2002), (b) 6*U* and 3*I* (April 2004), and (c) 7*CG* and 3*I* (March 2006).

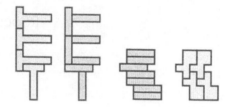

Figure 3. A few heptomino-tetromino solutions: (a) $7AE = 4I$, and (b) $7BH = 4I$

1. Cut out the five trominoes and three pentominoes. Without any overlapping, assemble them into two identical shapes.

2. Cut out the six trominoes and three hexominoes. Without any overlapping, assemble the nine pieces into two identical shapes using (i) a six piece group and a three piece group and (ii) a five piece group and a four piece group. (Note that any non-symmetric piece may be turned over.)

3. Cut out the seven trominoes and three heptominoes. Without any overlapping, assemble them into two identical shapes. (Note that any non-symmetric piece may be turned over.)

Other individuals have investigated pair-wise combinations of larger polyominoes up through the octominoes [7, 8]. A few of the more unusual examples from Resta are provided in Figure 3.

The comparison of larger *sets* of polyominoes has also been studied [7]. Figure 4 from Mireles shows the smallest area that can

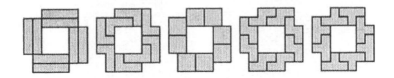

Figure 4. Smallest area solution for all five of the tetrominoes.

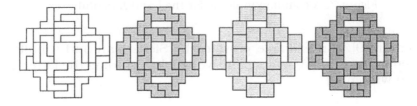

Figure 5. Smallest area solution for only the L, N, O, and T tetrominoes.

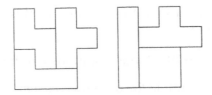

Figure 6. A solution for heterogeneous tetromino sets $4L + 4N + 4T = 4I + 4O + 4T$.

be tiled using all five of the tetrominoes. An interesting extension of this concept requires that another constraint be met, namely of allowing *only* a specified group of polyominoes to be used [10]. Figure 5 from Zucca illustrates the smallest area that can be tiled only by tetrominoes L, N, O, T and no others.

Liu has compared *heterogeneous* sets of polyominoes [5]. For example, Figure 6 shows an area tiled by the L, N, and T tetrominoes and also the I, O, and T tetrominoes. He refers to these comparisons as "Tetris Algebra" and has investigated all possible combinations using mixed sets of the five tetrominoes.

Rather than using polyominoes, this concept may also be applied to *other tile shapes* such as polyhexes and polyiamonds. The former have been studied by several others and reported by Ed

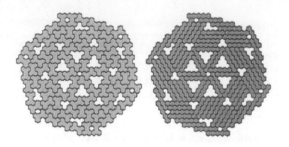

Figure 7. A common multiple for the I and X pentahexes.

Pegg Jr. [9]. An example provided by Sicherman and Cibulis is given in Figure 7. A comprehensive article regarding the latter has been written [6].

Open Questions and Challenges

Additional developments and extensions of this interesting topic include the following:

- Discover if any of the larger solutions can be reduced.

- Find proof or example for unknown cases.

- Explore combinations using even larger polyominoes.

- Consider three-dimensional shapes such as polycubes.

Bibliography

[1] U. Barbans, A. Cibulis, G. Lee, A. Liu, and B. Wainwright. "Polyomino Number Theory (II)." In *Mathematical Properties of Sequences and other Combinatorial Structures*, edited by J. S. No et. al., pp. 93–100. Dordrecht: Kluwer Academic Publishers, 2003.

[2] U. Barbans, A. Cibulis, G. Lee, A. Liu, and R. Wainwright. "Polyomino Number Theory (III)." In *Tribute to a Mathemagician*, edited by B. Cipra et. al., pp. 131–136. Wellesley, MA: A K Peters, 2005.

[3] A. Cibulis, A. Liu, and B. Wainwright. "Polyomino Number Theory (I)." *Crux Mathematicorum* 28 (2002), 147–150.

[4] S. W. Golomb. "Normed Division Domains." *American Mathematical Monthly* 88 (1981), 680–686.

[5] A. Liu, et al. "Tetris Algebra." Unpublished paper, 2006.

[6] M. Lukjanska, G. Sicherman, and A. Liu. "Polyiamond Number Theory." *Journal of Recreational Mathematics* 33:1 (2004–2005), 39–47.

[7] J. Mireles. "Poly^{2}ominoes." Available at http://www.geocities.com/jorgeluismireles/polypolyominoes/, 2003.

[8] G. Resta. "Polypolyominoes." Available at http://www.imc.pi.cnr.it/resta/Poly/, 2004.

[9] G. Sicherman and A. Cibulis. "Pentahex I and X." Available at http://www.mathpuzzle.com/, 2008.

[10] L. Zuca. "Tetrominoes Challenge." Available at http://www.geocities.com/liviozuc/, 2004.